6|17

The Truth about Language

The Truth about Language

What It Is and Where It Came From

Michael C. Corballis

THE UNIVERSITY OF CHICAGO PRESS

CHICAGO AND LONDON

The University of Chicago Press, Chicago 60637
The University of Chicago Press, Ltd., London
© 2017 by Michael C. Corballis
All rights reserved. Published 2017.
Printed in the United States of America

26 25 24 23 22 21 20 19 18 17   1 2 3 4 5

ISBN-13: 978-0-226-28719-5 (cloth)
ISBN-13: 978-0-226-28722-5 (e-book)
DOI: 10.7208/chicago/9780226287225.001.0001

Library of Congress Cataloging-in-Publication Data
Names: Corballis, Michael C., author.
Title: The truth about language : what it is and where it came from /
    Michael C. Corballis.
Description: Chicago ; London : The University of Chicago Press, 2017. |
    Includes bibliographical references and index.
Identifiers: LCCN 2016027786 | ISBN 9780226287195 (cloth : alk. paper) |
    ISBN 9780226287225 (e-book)
Subjects: LCSH: Language and languages—Origin. | Language and
    languages—Philosophy. | Thought and thinking.
Classification: LCC P116 .C6714 2017 | DDC 401—dc23
    LC record available at https://lccn.loc.gov/2016027786

⊚ This paper meets the requirements of ANSI/NISO Z39.48–1992 (Permanence of Paper).

Language is my whore, my mistress, my wife, my pen-friend, my check-out girl. Language is a complimentary moist lemon-scented cleansing square or handy freshen-up wipette. Language is the breath of God, the dew on a fresh apple, it's the soft rain of dust that falls into a shaft of morning sun when you pull from an old book-shelf a forgotten volume of erotic diaries; language is the faint scent of urine on a pair of boxer shorts, it's a half-remembered childhood birthday party, a creak on the stair, a spluttering match held to a frosted pane, the warm wet, trusting touch of a leaking nappy, the hulk of a charred Panzer, the underside of a granite boulder, the first downy growth on the upper lip of a Mediterranean girl, cobwebs long since overrun by an old Wellington boot.

Stephen Fry, from *A Bit of Fry and Laurie*

Contents

Preface ix

Part One: Background to the Problem

1 The Rubicon 3
2 Language as Miracle 24
3 Language and Natural Selection 40

Part Two: The Mental Prerequisites

4 Thinking without Language 57
5 Mind Reading 81
6 Stories 101

Part Three: Constructing Language

7 Hands On to Language 123
8 Finding Voice 147
9 How Language Is Structured 172
10 Over the Rubicon 189

Notes 205
Bibliography 231
Index 251

Preface

Language is the elephant in the room, the jewel in the crown, the ghost in the machine. It is perhaps the ultimate challenge for the social and biological sciences, since no one really understands how it works, yet, barring disease or misadventure, we all possess it. Without language there would be no stories, no religion, no science, no history. Some would say no consciousness—wrongly, I think, but that's a story for later. And yet we are the only species that can communicate in that open-ended way that we like to call language, filling our daily lives with talk and gossip, our libraries with books, our televisions with soap operas and excitable sports commentators, our parliaments with vacuous bickering and self-important posturing, our computer screens with downloads of variable authenticity, our lecture halls with bespectacled wisdom—not to mention the twittering of our smartphones.

Strangely, though, we seem to take language for granted, a gift bestowed on us for the privilege of being human. Of course other animals do communicate, but their communications have nothing approaching the sheer vastness of human language, its extraordinary power to evoke, explain, persuade, recount—and of course bullshit. Animals can of course communicate, conveying pain or emotion, but their apparent inability to tell us their thoughts, ideas, memories, or plans somehow seems to absolve us from guilt over the ways we exploit them. Perhaps it has seemed better not to question how we came to possess language, but rather to assume that it's simply a mark of our superiority, placing us closer to the angels than to the apes.

In any event, language seems so different from any other form of communication, whether the chirruping of birds or the chattering of monkeys, that it almost defies explanation. Throughout history, therefore, there has been a strong temptation to suppose that it was simply bestowed on us by some deity or maybe was an outcome of some fluke of nature—a mutation, perhaps, or a property emerging from an expanded brain. From a Darwinian perspective, though, this won't do. The challenge is to place language, like any other complex faculty or organ, into the context of natural selection.

Of course some have tried. The esteemed behaviorist B. F. Skinner sought to explain language in terms of basic behavioral principles derived from work with animals, principally pigeons. This work, described in his 1957 book *Verbal Behavior*, did not really propose an evolutionary scenario but simply set language in the context of animal behavior, requiring no special discontinuity between ourselves and other animals. Skinner's work also implicitly recognized that language should not be identified with speech but is rather a form of behavior—a recognition that resonates with one of the themes of this book. But Skinner's legacy has not really lasted; his influence was thwarted by the publication in the same year of another much slimmer volume by the linguist Noam Chomsky, *Syntactic Structures*. Chomsky's book effectively denied that language could be understood in terms of associative learning and reopened the chasm between humans and even our closest nonhuman relatives, the great apes. These and ensuing events are covered in the pages that follow.

The divide between those who favor a progressive, Darwinian account and those who believe language to have been the result of some sudden and dramatic change remains as large as ever. And of course I am not the first to attempt an account of language evolution in terms of natural selection. There have been intermittent attempts through history, often opposed by the church, and also a flurry of recent accounts, opposed not so much by religious authorities as by those who think that the gap between humans and other animals is simply too great to have been breached by the incremental steps of evolution. This issue also plays out in more detail in the pages of this book.

< x >

I am grateful for discussion with many who have some broad agreement with the approach I have taken, including Michael Arbib, Christina Behme, Richard Byrne, Nicola Clayton, Francesco Ferretti, Russell Gray, James Hurford, Giacomo Rizzolatti, Kim Sterelny, the late William Stokoe, and Sherman Wilcox. Many of my students have been kind enough to agree with my views on language evolution, but then I suppose they would, wouldn't they? On the other side of the coin I have (mostly) enjoyed sparring with Chomsky himself, as well as with Tecumseh Fitch, Mark Hauser, Adam Kendon, Robert Seyfarth, my good friend Thomas Suddendorf, and Ian Tattersall. I don't suppose I will persuade those to whom the gap is unbreachable in Darwinian terms, but perhaps I can at least contribute to what seems to me to be a sea change in our understanding of language and its evolution.

On a more personal level, I am grateful to Barbara Corballis for support, and also to my brilliant sons Paul and Tim and to their equally brilliant daughters Simone, daughter to Paul and Theresa, and Lena and Natasha, twin daughters to Tim and Ingrid. The three girls are all currently aged seven, and already more eloquent than their grandfather.

< xi >

PART ONE

Background to the Problem

Over a decade ago, Morten Christiansen and Simon Kirby introduced an edited collection of articles on the evolution of language with the chapter title "Language Evolution: The Hardest Problem in Science?" It was framed as a question but may indeed be true as a statement. In this book I attempt a solution—one that is in part speculative but based where possible on facts. Part 1 sets the background and has three chapters.

Chapter 1 describes some of the properties of language that make it seem so intractable. It opens with a quote from a prominent nineteenth-century philologist writing, as many did at the time, in protest against Darwin's theory of evolution. Language, he thought, was the one obstacle to the idea that human behavior could have arisen through natural selection. Language indeed seems to be unique to our species, and to have properties not easily discerned either in other aspects of human thinking or in the behaviors of our closest nonhuman relatives.

In chapter 2 I outline how the apparent uniqueness and complexity of language have led to the view that language must have been the result of some miracle, whether a gift from the deity, a fortunate genetic mutation, or simply a byproduct of having a large and complicated brain. Prominent among those who argue that language emerged in our species in a single step is Noam Chomsky, the foremost linguist of our time, and his views are supported by many contemporary linguists and anthropologists.

< 1 >

Chapter 3 then provides a background to the understanding of language as a product of gradual evolution. In a post-Chomskian era, some theorists are edging toward a Darwinian account, although there is as yet little agreement as to the main steps.

The stage is then set for a more detailed examination of how language might indeed have evolved.

1

THE RUBICON

Where, then, is the difference between brute and man? What is it that man can do, and of which we find no signs, no rudiments, in the whole brute world? I answer without hesitation: the one great barrier between the brute and man is Language. Man speaks, and no brute has ever uttered a word. Language is our Rubicon, and no brute will dare cross it.

So declared Friedrich Max Müller (1823–1900), professor of philology at the University of Oxford, in a lecture on the science of language delivered in 1861. Müller was protesting against Charles Darwin's famous treatise *On the Origin of Species*, which had been published just two years earlier.[1]

The essence of Darwin's theory of evolution is natural selection, the process by which biological traits become more or less common in a population. This in turn depends on natural variation between organisms, so that variants with higher rates of reproduction become more populous. The nature of this "selection" is such that it has no purpose or direction. Because the variation is small, evolution works slowly and in small increments. Darwin wrote without knowing anything about genes or DNA, but we now know that genes are subject to mutations, creating the variations upon which natural selection operates.

To Müller, then, the difference between language and animals' communication was simply too profound to have come about through incremental tweaking—too wide a Rubicon for evolution, with its

< 3 >

mincing little steps, to cross. And language is widely considered the commodity that most clearly defines us as human. Barring exceptional circumstances, we all acquire it. That in itself is not extraordinary, because we also learn to walk, just as birds learn to fly. Language, though, seems different, in that it is complicated and allows a freedom of expression far beyond that available even to our closest nonhuman relatives, chimpanzees and bonobos. Even linguists don't yet fully understand the rules by which we generate sentences or tell coherent stories. In contrast, the "brutes" that Müller disparages communicate in very limited and stereotyped ways, at least if we consider vocal communication. I shall argue later, though, that the seeds for a more flexible form of communication lie in the hands rather than the voice.

The most dominant languages in the modern world are English and Chinese, which are vastly different from one another. Chinese has the largest number of native speakers, but English takes the lead if you include those who speak it as a second language. Chinese is complicated by the fact that there are several versions; these are generally regarded as dialects of a common language but may in fact be as diverse as the Romance languages. Nevertheless the great majority of Chinese people, some 960 million, speak Mandarin Chinese as their native language, and that alone probably puts Chinese in the ascendancy—ahead of Spanish with about 400 million. Ironically, English and Chinese are among the most difficult languages for nonnative speakers to learn. Chinese is a tonal language, and getting the tone wrong can lead to misunderstanding; you may think you're saying *jī*, meaning "chicken," but a false note yields *jì*, meaning "whore." English has consonant clusters that are awkward for non-English speakers, as in *street* or *exempts*, and boasts some twenty different vowel sounds, as in *par, pear, peer, pipe, poor, power, purr, pull, poop, puke, pin, pan, pain, pen, pawn, pun, point, posh, pose,* and *parade.* Spanish, in contrast has only five vowel sounds.[2]

In spite of the oppressive dominance of English and Chinese, at least six thousand different languages are spoken around the globe, each more or less unintelligible to the rest. An extreme example is the Pacific archipelago of Vanuatu, with an area of only about 4,379 square miles, which is host to over one hundred different languages.[3] Sometimes we have difficulty understanding even those who supposedly speak the same language; George Bernard Shaw once remarked that "England and America are two countries separated by the same language." He might also have had Scotland in mind, because the English dialogue in the 1996 movie *Trainspotting*, set in Scotland, required subtitles when shown in the United States. Language is deeply cultural, and serves to exclude outsiders as much as to bind insiders together. As the title of Robert Lane Greene's recent book puts it, *You Are What You Speak*.

But we shouldn't be complacent, because it has been estimated that over twenty-four hundred of the world's languages are in danger of disappearing.[4] Around a quarter of living languages have fewer than one thousand speakers, and many languages spoken by local communities are being replaced by dominant regional, national, and international languages. Mark Turin refers to the loss of languages as "linguicide."[5]

Sign languages too are diverse, in spite of the fact that signs generally originate as mimed representations of objects or actions. In the course of time, these representations become stylized—or conventionalized, to use the technical term—and so lose much if not all of their pictorial or action-based character. Sign languages are typically invented anew by different deaf communities, and different sign languages are just about as mutually unintelligible as are different spoken languages.

In spite of the extraordinary differences between the languages of the world, though, it seems safe to assume that any person can learn any language, provided they start early in life. This suggests that language is as much biological as cultural—the capacity to learn it is biological, but the form it takes depends on culture. There remains a question as to whether this biological capacity for language is specific to language itself or comes about because we humans are smart and inventive in general ways. Nevertheless, as far as we know we are the only species with that capacity. Our closest nonhuman relatives are chimpanzees and bonobos, with whom we share a common ancestry dating back six or seven million years. In geological time this is really just an eye-blink away from the present, and it has also been estimated that we share some 99 percent of our genes with these oddly humanlike animals.[6] Attempts to teach them language, though, have failed rather miserably. To be sure, a few have been trained to make simple requests using a form of sign language rather than speech, but there are few if any glimmerings of gossip, reminiscence, observations about the world, storytelling, or explanations of how things work. Parrots can learn to utter words and even give answers to simple questions, but they too do not use language in the flexible way that we humans do. They can be agreeable and friendly companions, but they are not really candidates for a conversation, and they cannot tell us what it's like to be a parrot. Language-wise, we humans seem to be alone in the world—and possibly in the universe.[7]

Language is not only uniquely human—it is also universally so. In every part of the world, people speak (or sign) to one another, although there are of course a few interesting exceptions. Children isolated from human contact do not learn to speak properly (some such cases are the stuff of legend more than of fact). Reports of so-called wild children brought up by animals, including wolves and bears, have long featured in folklore and have formed the basis of such fictional characters as Rudyard Kipling's Mowgli, J. M. Barrie's Peter Pan, or Edgar Rice Burroughs's Tarzan. Whether there are truly instances of human children raised by animals is doubtful.

The celebrated case of Amala and Kamala, two girls reportedly discovered by missionaries in a forest in India and said to have been raised by wolves, turned out to be a ruse to attract funds for the orphanage in which they were eventually placed. The best-documented case of a child deprived of a normal social environment is Genie, a Californian girl who was isolated by her family from infancy until the age of thirteen. When she was then discovered, she attracted great interest from psychologists and linguists, and strenuous efforts were made to teach her to speak. She did develop some ability to communicate by vocalizing and gesturing, and even by drawing, but she never ac-

< 6 >

quired normal grammatical speech.[8] The best she could manage was a kind of telegraphese, a sort of "me Tarzan you Jane" level of speaking. Such examples have led to the idea of a "critical period" for the learning of language; once you pass puberty, it seems, the game is all but over.

What this suggests is that acquiring a first language can take place only when the brain is itself developing. Of course people do learn second languages as adults, but it can be a hard slog, and it seems impossible to get rid of a foreign accent. This is in marked contrast to the effortless way in which young children learn languages. Learning a second language as an adult, moreover, is not the same as learning a first language, because you can use the first language as the scaffold on which to build the second. And because the brain is at its most plastic and impressionable while growing, the secret of language may well lie partly in the prolonged period of growth that our large brains undergo. Most of this growth occurs after birth, so that the developing brain is exposed to the world outside of the womb and can be shaped by the sights and sounds that the world inflicts on us. Compared to monkeys and apes, we humans are born prematurely and spend a longer time to reach maturity. It has been said that in terms of the general pattern followed by other primates, human babies should be born at eighteen months of gestation, not nine. But birth is difficult enough as it is without having to wait another nine months; even I, as a hapless male, can appreciate that.

Early birth was probably driven by the fact that our species, unlike the other apes, elected to stand and walk on two legs rather than four—to reverse the slogan of the rampant pigs in George Orwell's *Animal Farm*, "two legs good, four legs bad!" This in turn restricted the size of the birth canal, so our kids need to be born before they grow too large. Even so, birth is difficult, as any mother can attest, but the tradeoff is that human babies are exposed to the postwomb environment while their brains are still immature and ready to be shaped by the social and physical environments into which they are born. Our persistent two-legged stance is in many ways an impediment, giving rise to back and neck problems, hemorrhoids, hernias, and of course

< 7 >

the excessive pain of giving birth. Bipedalism, one might say, is a pain in the ass. But one far-reaching advantage is that it extends the period of growth outside the womb, allowing the brain to grow and adapt while exposed to the sights and sounds of the world.

We are bathed in language from very early in life. Even at one day old, babies can tell their mother's voice from that of a stranger,[9] suggesting that tuning in to the mother's voice takes place in the womb. Not for nothing do we speak of a mother tongue. We should not forget, too, that language is not wholly a matter of voice, because we gesture and point in the course of normal discourse—and of course sign language is entirely a matter of gesture. Ultrasound recordings show that fetuses in the first trimester move their arms, and most move the right arm more than the left.[10] In the second trimester they suck their thumbs, and again it's more often the right thumb than the left.[11] These asymmetries may well set the stage for the fact that most of us are right-handed and have language controlled by the left side of the brain.

But it's only when they emerge into the light and bustle of day that babies can begin to associate sounds or gestures with the diversity of what they can see, touch, and hear. The babbling of babies in the first year begins to take on some of the characteristics of the language they are exposed to, and between ages one and two pointing plays the major role.[12] The very helplessness of human infants also adds to the impact of language, because it brings them into closer contact with caregivers. There's nothing like the sight of a newborn baby to bring out infantile behavior in otherwise serious and responsible adults, as their language deteriorates into baby talk with simplified words, cooing sounds, clucks, and goos. This is known as motherese—although in a politically correct world it is now more often called parentese. Even dads can cluck and goo.

So it is that we mold our babies' babbles into words. We know too that manual and facial gestures play a role in helping infants learn spoken as well as signed language. In the early years, at least, pointing is essential for learning the names of things, even if the names themselves consist of signs rather than spoken words. Young babies often

< 8 >

point in order to share attention, as if to say "Look at that!" whereas chimpanzees point mainly to make requests, as if to say "Gimme that!" Shared attention through pointing is one of the first indications of an inborn disposition for language.[13]

Although it depends on early experience, language has a robustness that defies at least some forms of disability or disadvantage. As I have already mentioned, communities of deaf people, denied normal speech, spontaneously develop signed languages, carried out silently with movements of the hands and face. Indeed I shall argue later in this book that language evolved from manual signs rather than from animal calls. Language is normally lodged in the left side of the brain, but if the left side is damaged early in childhood, or even removed, the right side can take over with little impediment. Our very brains seem to burst with the desire for expression. It takes extreme circumstances, such as those suffered by Genie, to prevent language from developing normally.

Regardless of the language or languages we speak or sign, we follow rules for how to string words or hand movements together to form meaningful content. The way we do this is complex, and linguists have still not fully explained the rules that govern it, whether they are specific to individual languages or apply generally across languages. It is a singular fact that speakers of any given language know the rules at some intuitive level, so they can generally tell whether a given utterance is grammatical or not, but they cannot tell you exactly what those rules are.

The rules need not conform to textbook definition or what high-school teachers tried to instill in reluctant students. Slang and street talk also follow rules. People seldom diverge from the language of their group, and they even switch depending on whom they're talking to. Teenagers speak to other teenagers differently from how they talk to their parents or teachers. Whatever they are, though, the rules operate in open-ended fashion, such that there is in principle no limit to the number of things we can say or sign. Noam Chomsky referred to language as possessing the property of "discrete infinity." That is, we have a finite number of discrete sounds or signs,

< 9 >

but these can be combined in a potentially infinite number of ways to create new meanings. We can produce sentences we have never previously uttered and understand sentences we have never heard before—provided of course they are made up of words put together in ways that we are familiar with.[14]

My favorite example occurred when I called in to a publishing house in the south of England a few years ago. I was greeted at the door by the publisher himself, who said, "We are having a bit of a crisis here. Ribena is trickling down the chandeliers." The words *ribena*, *trickling*, and *chandeliers* were familiar to me, but I had never before heard them in that particular combination; still, I understood the publisher's predicament. Ribena is a drink made from black currants and is high in Vitamin C content; for some time it was delivered free to English schoolchildren. The publisher and his concerned staff had initially thought that a red substance trickling from their chandeliers was blood, suggesting that some foul deed had taken place upstairs. It transpired that there was a nursery school upstairs, and one of the little girls had thought it fun to tip her ribena onto the floor instead of into her mouth. As they do.

A more famous example was coined by the philosopher Alfred North Whitehead, in conversation with Burrhus Frederick (B. F.) Skinner, the well-known behavioral psychologist. Skinner was extolling the power of behaviorism to explain what people do and even what they say, so there is no need to appeal to mental processes. Listening to this, Whitehead was moved to utter the sentence "No black scorpion is falling upon this table" and ask Skinner to explain why he said that. This conversation took place in 1934, and it was not until 1957, in an appendix to his book *Verbal Behavior*, that Skinner attempted an answer. For a behaviorist dismissive of psychoanalysis, Skinner gave a curiously Freudian interpretation. He proposed that Whitehead was unconsciously likening behaviorism to a black scorpion and declaring that it would have no part in his understanding of the human mind.

Ironically, though, 1957 was also the year in which Noam Chom-

< 10 >

sky published his book *Syntactic Structures*, which presented a view of language totally opposed to a behaviorist account. Two years later, Chomsky made explicit his objection to behaviorism in a scathing review of *Verbal Behavior*.[15] Where Skinner regarded language as vocal behavior emitted by speakers and reinforced by the language community, Chomsky proposed that language must depend on innate rules to govern the formation of sentences. Reinforcement of sequences simply could not explain the sheer novelty and diversity—the "discrete infinity"—of natural language.

Our ability to generate sentences of seemingly endless variety arises from combinations rather than from the simple accumulation of elements. There are 311,875,200 different poker hands of five cards that be dealt from the full deck of 52. This illustrates how vast, if not infinite, numbers of combinations can arise from relatively small vocabularies. This example is a bit misleading, though, because not all combinations of words are meaningful, but our deck of words is much larger than 52—a college-educated person may have a vocabulary of some 50,000 words.[16] To be slightly more realistic, suppose that we distinguish between words corresponding to objects and words corresponding to actions, so we can compose utterances like "man walks" or "elephant dances." Let's suppose that our deck contains 40 object words and 12 action words. We can then compose 40×12, or 480, utterances—still a considerable advance over the 52 elements, although this too produces some utterances that scarcely match reality, such as "tree laughs" or "butter pontificates." But then we can add other elements, such as the victim of an action, as in "man bites dog," or add another object as part of a transaction, as in "girl gives dog bone." We can then start adding words to describe qualities and other paraphernalia to indicate place, time, and so on, and the possible combinations multiply by orders of magnitude, as in "Yesterday, that lovely young girl generously gave my dyspeptic dog a disgusting old bone."

In his short story "The Library of Babel" the Argentinian writer Jorge Luis Borges took the idea of discrete infinity to a limit—

< 11 >

although technically not actually reaching infinity. The narrator in the story inhabits a universe consisting of a huge beehivelike expanse of hexagonal rooms, each with four walls of bookshelves. The books themselves contain all possible combinations of the 25 characters in Spanish—22 letters, a period, a comma, and a dash. Although the vast majority of the books are gibberish, they must also include all the books ever written, and all that will be written, including this one. The library also includes every possible book, so that the only books that are excluded are the impossible ones—the narrator notes, for instance, that "no book can be a ladder, although no doubt there are books which discuss and demonstrate and negate this possibility and others whose structure corresponds to that of a ladder."[17]

The number of books in that library is at least $25^{1,312,000}$. This amounts to about $1.956 \times 10^{1,834,097}$, which is, oh, a lot, when you consider that current estimates of the number of atoms in the universe is only 10^{80} or so. Even so, the number is still not infinite, but about as close as one can imagine.

Recursion

But there's more. We can add to the stretch toward infinity by adding recursive principles so that combinations can be embedded in combinations. "The House That Jack Built" provides a favorite, if perhaps overworked, example:

> This is the farmer sowing his corn
> that kept the cock that crowed in the morn
> that woke the priest all shaven and shorn
> that married the man all tattered and torn
> that kissed the maiden all forlorn
> that milked the cow with the crumpled horn
> that worried the dog
> that chased the cat
> that killed the rat

< 12 >

that ate the malt
that lay in the house
that Jack built.

In principle this could go on indefinitely. You could insert *the chap who was nobly born who accosted* after the third word. Or you might add *John said that Mary thought that Fred claimed that* at the beginning. There are of course psychological limits to how much embedding you can take, because you might simply forget how the sentence began by the time you reach the end, or fail to remember the full cast.

In "The House That Jack Built" clauses are added to the right. This is called right-embedding. Much more psychologically taxing is so-called center-embedding, where clauses are inserted in the middle of clauses. We can cope with a single embedded clauses, as in

The malt that the rat ate lay in the house that Jack built.

But it becomes progressively more difficult as we add further embedded clauses:

The malt [that the rat (that the cat killed) ate] lay in the house that Jack built.

Or worse:

The malt [that the rat (that the cat {that the dog chased} killed) ate] lay in the house that Jack built

I added brackets in the last two examples that may help you see the embeddings, but even so they're increasingly difficult to unpack. Center-embedding is difficult because words to be linked are separated by the embedded clauses; in the last example above, it was the malt that lay in the house, but the words *malt* and *lay* are separated by twelve words. In holding the word *malt* in mind in order to hear what happened to it, one must also deal with the separations between

< 13 >

rat and *ate* and between *cat* and *killed*. The mind boggles—but "The Library of Babel" still copes.

In spoken English, single embedded clauses are common enough, but double embeddings are rare and triple embeddings (as in the last example) virtually nonexistent. Center-embeddings are more common in written language than in spoken language, perhaps because when language is written you can keep it in front of you indefinitely while you try to figure out the meaning. The Finnish linguist Fred Karlsson examined a huge corpus of writings in several European languages and found only thirteen instances of triple center-embeddings.[18]

Although the embeddings of phrases within phrases adds to the complexity of language and increases the number of sentences that can be derived from a finite set of elements, its importance may well be exaggerated. Some languages do not seem to incorporate recursive embedding at all. Two such examples are the languages of the Pirahã, a remote Amazonian community, and the Iatmul of New Guinea.[19] Linguists commonly argue that the constraints that prevent us from using multiple embeddings, or from generating more than a rather limited corpus of utterances, are psychological rather than linguistic. The linguistic rules that underlie our language faculty can create utterances that are potentially, if not actually, unbounded in potential length and variety. These rules are as pure and beautiful as mathematics. But perhaps they are not real. The rules themselves may be more limited than linguists like to think.

Universal Grammar?

Humans seem uniquely and universally disposed to acquire language, whether spoken or signed. Moreover, children readily learn whatever language they are exposed to. A child from the highlands of New Guinea, say, if raised in Boston, will readily acquire Bostonian English. Such observations have led Noam Chomsky, the promi-

< 14 >

nent linguist of the past half-century, to conclude that language depends on a biologically determined instinct, which he calls "universal grammar."[20]

Given that there are some six thousand languages in the world, the challenge is to specify what they have in common or what rules of universal grammar might apply across all languages. Chomsky himself seems to have determinedly ignored the vast variations in human languages, at one stage remarking: "I have not hesitated to propose a general principle of linguistic structure on the basis of observation of a single language. The inference is legitimate, on the assumption that humans are not specifically adapted to learn one rather than another human language."[21] This insistence that a single language can reveal all we need to know about the general nature of language may be the Achilles' heel of any notion of universal grammar. Any universality may lie not so much in language itself as in the common ways we parse the world, as I shall suggest later in this book. For instance, all languages probably have ways to refer to objects and actions, but that could be because our worlds are largely composed of these elements, and if we didn't have innate ways to refer to them we'd no doubt soon invent them. Of course our parsing of the world into objects and actions in turn depends on our biological makeup; mosquitoes or fish may well see their worlds quite differently.

The way in which we map the events in the world onto words also varies between cultures. In English we generally think of words as belonging in different categories, such as nouns, verbs, adjectives. But some categories are missing in some languages. One example is articles. In English, for example, the definite article *the* and the indefinite article *a* are used to distinguish between entities that are specified from those that are not: *the elephant* refers to a particular elephant under discussion, but *an elephant* refers to some unspecified elephant—perhaps one that just happened to wander onto a basketball court. But there are no articles in Russian or Chinese. The use of tense to mark verbs according to when some activity or situation occurred also occurs in some languages but not in others. In languages

< 15 >

like Latin and Italian, this is done by altering the verb itself. Thus in Latin *video* means "I see," *videbam* means "I was seeing," *videbo* means "I will see," and so on—and on. English seems simpler in that we have only four basic forms of regular verbs (*argue, argues, argued, arguing*), but we change tense or mood largely by adding auxiliaries in often complex ways ("they might have been arguing"). But Chinese has no tenses at all and uses other devices to indicate when something happened or will happen.[22]

Across different languages it is not even clear which words belong in which category. Nicholas Evans gives several examples. The concept of *paternal aunt*, a noun phrase in English, is expressed by a verb in the Australian aboriginal language Ilgar, and the expression of *love* is simply a suffix in the South American language Tiriyo. Words generally hold single meanings, but some words are hosts to several meanings. For example, in the language of the Seneca people, an indigenous group in North America, words describing events also carry the idea of the participant or participants in the event. Thus the Seneca verb *wa'e:yeh* includes not only the idea of an event of "waking up" but also the idea of the person who woke up, in this case a female. The English equivalent would be *she woke up*, where the participant is captured in the separate word *she*. In Seneca it is conveyed by the *e* sound in the second syllable of *wa'e:yeh*, but this sound by itself does not convey the notion of the person who woke up in the way that the word *she* does in English. Unlike English, Seneca is *holophrastic* in that words can stand for events in a holistic manner, where the component parts of the event are inseparable.

In some languages quite complex concepts may be packed into a single word. Turkish is one extreme example of a so-called agglutinative language, where there are said to be over two million forms of each verb. The different forms represent not only the tense but also the subject, object, and indirect objects of the verb—and a lot else besides. Nouns also take multiple forms. The word *arabamizdakilerinle* means "with those, in our car, that are being possessed by you." In some languages, entire sentences are packed into a single word. Nicholas Evans

< 16 >

and Stephen Levinson give the examples of *Ęskakhǫna'tàyęthwahs* from the Cayuga of North America, which means "I will plant potatoes for them again," and *abanyawoihwarrgahmarneganjginjeng* from the Northern Australian language Bininj Gun-wok, and means "I cooked the wrong meat for them again."[23]

Comparing languages across differing cultures suggests an inverse relation between the complexity of grammar and the complexity of culture; the simpler the culture in material terms, the more complex the grammar. Mark Turin notes that colonial-era anthropologists set out to show that indigenous peoples were at a lower stage of evolutionary development than the imperial Western peoples, but linguistic evidence showed the languages of supposedly primitive peoples to have surprisingly complex grammar. He writes: "Linguists were returning from the field with accounts of extremely complex verbal agreement systems, huge numbers of numeral classifiers, scores of different pronouns and nouns, and incredible lexical variation for terms that were simple in English. Such languages appeared to be untranslatable."[24]

Turin writes that Thangmi, an unwritten Tibeto-Burman language spoken by around thirty thousand people in north-central Nepal and in part of West Bengal in India, often has several words to express meanings conveyed by a single word in English or French. For example, Thangmi has four words, with completely different stems, to express the notion "to come" in the contexts of coming up a hill (*wangsa*), down a slope (*jusa*), from the same level or around an obstacle (*kyelsa*), or from an unknown or unspecified direction (*rasa*). In English, in contrast, we can use a single verb to express the idea of coming and then qualify it with additional words; thus the different meanings are created by combining words rather than by packing meaning into individual words. This move toward combinatorial structure may be heightened in complex cultures where there are simply more objects and concepts to name, so the pressure to coin new words may have been offset by restricting the different forms that individual words can take. The industrial and computer

< 17 >

revolutions, to name two major cultural forces, created vast numbers of new concepts for which we need words. Even sentences appear to have become simpler in the world of Twitter.

Another influence may relate to the forms of language itself. Sign languages tend to be agglutinative, probably taking advantage of the visual system's capacity to process different aspects of the world simultaneously.[25] The auditory system, in contrast, is better adapted to processing rapid sequences of information. The persistence of agglutinative languages may conceivably reflect the origins of language in gesture and pantomime, with the agglutinative structure tending to break down as speech, with its relentless sequential structure, grew more dominant. I explore this possibility in more detail in chapter 7.

Based on the remarkable differences between languages, not only in the structure of words but also in the ways they are combined into grammatical sequences, Nicholas Evans and Stephen Levinson conclude that "the emperor of Universal Grammar has no clothes."[26] Another expert on language and its evolution, Michael Tomasello, once remarked similarly that "universal grammar is dead."[27]

Language Templates

One problem with theories of grammar has to do with the idea of discrete infinity itself. As suggested earlier, although we may be able in principle to generate sentences of unlimited length and variety, our actual ability is really rather limited. Even the novelist Henry James, famous for the long sentences that perfused much of his writing, had his limits. According to the *Guinness Book of World Records*, the longest sentence in literature comes from William Faulkner's novel *Absalom Absalom!* and contains 1,288 words. Literature gives a rather inflated estimate of sentence length in any case, because one can hold the sentence on the open page while one tries to figure it out. Even in literature, though, it seems that sentences become shorter over time, at least in English. Edmund Spenser, writing at the end of the sixteenth century, averaged 50 words a sentence, whereas Thomas Macaulay

< 18 >

in the middle of the nineteenth century averaged just 23. The trend may have continued.[28] Brian Hayes happened upon the syllabus for an English course at Northeastern University that begins "Choose a big Victorian novel not to read."[29]

In normal conversation we are much more restricted. And although recursion is one of the major mechanisms underlying discrete infinity, we have seen that our ability to embed phrases within phrases is actually very limited, and some languages do not seem to exhibit this facility at all. The paucity of everyday language relative to the potential for discrete infinity can be attributed to our limited memory, restricted powers of attention, or inability to plan more than a few steps ahead. Some have suggested, though, that the rules of language are themselves more limited than the somewhat idealized rules of grammar imply.

Much of our language actually seems to rely simply on memorized stock phrases that we slip into our conversations, rather than on the application of rules. Alison Wray gives some examples:

> Fancy seeing you here.
> Watch where you're going.
> You'll never guess what happened.
> How dare you![30]

Much of conversation proceeds through clichés of this sort.

The English comedian and raconteur Stephen Fry, in conversation with Hugh Laurie, gives more examples:

> I love you.
> Don't go in there.
> You have no right to say that.
> Shut up.
> I'm hungry.
> That hurt.
> Why should I?
> It's not my fault.

< 19 >

Help!

Marjorie is dead.[31]

We might go further and suppose that we use memorized stock phrases to create new ones, which is simpler and more immediate than generating sentences from basic rules. James Hurford gives the example of the idiom *having a chip on one's shoulder*.[32] One can then substitute new words within this frame, as in *having milk in one's coffee, having butter in one's fridge, having money in the bank*—and there are even other idioms based on the same structure, like *having a bee in one's bonnet*. Much of the generativity of language may then derive from adapting stock expressions by altering the key words, or changing person or tense, as in *she had a chip on her shoulder*. The most extreme example of this approach is that of the Dutch computational linguist Rens Bod, who suggests that speakers store all of the sentences they have heard and then adapt them to new situations.[33] This would surely stretch memory capacity too far and create enormous redundancy, but it does reinforce the point that much of our ability to produce sentences depends on stored information, rather than the use of the rather idealized rules proposed in much grammatical theory. It also effectively denies universal grammar.

Bod's theory may perhaps impose impossible demands on memory, but much of our ability to produce sentences is indeed probably based on our knowledge of phrases and sentences we have heard, rather than on the application of innate rules. This has led to what has been termed the "data oriented parsing model" (DOP) of sentence production, which holds that people learn the structure of sentences, not from innately given linguistic rules, but from their linguistic experience. They can then select an appropriate structure from overlaps with sentences previously experienced. It's perhaps not so much that people commit actual sentences to memory as that they learn implicitly how sentences work—how they sound or look, and how they are structured. Bod has also adapted this approach to account for how children build up the experience to construct sentences.[34]

Perhaps because of the logjam of theories of grammar, the 1990s in fact saw something of a general swing toward explanations of grammar in terms of what we learn, rather than what we're born with. Morten Christiansen and Nick Chater have developed theories of language learning based on connectionism—the idea that the operations of the mind can be simulated with vast numbers of connections between elements that can be modified through experience. This model even simulates recursion, supposedly a key ingredient of universal grammar. As suggested above, our capacity for embedding is limited, and Christiansen and Chater's model more realistically places recursion within human limits rather than supposing it to depend on a process that is theoretically unlimited. They write as follows: "This work suggests a novel explanation of people's limited recursive performance, without assuming the existence of a mentally represented competence grammar allowing unbounded recursion."[35]

These developments therefore seem to be edging away from Chomskian orthodoxy. It is clear that we house not only a vast store of words but also a compendium of phrases and that we can use them to generate new utterances with little or no reference to underlying rules. This is the essence of what has come to be known as *construction grammar*,[36] which is also part of an approach known as *cognitive linguistics*,[37] whereby language is understood in terms of the general ways in which people think rather than in terms specific to language itself. Language then becomes a product of learning rather than of innate rules, although the human capacity to learn, make generalizations, and invent variations pushes language to a level of complexity well beyond the communication systems developed by any other species.

The idea that language competence is learned rather than innate takes us back to 1957 and the vastly different views of language on display in Skinner's *Verbal Behavior* and Chomsky's *Syntactic Structures*. In the ensuing decades, theories of grammar mushroomed, to the point that in 1982 James D. McCawley published a book titled *Thirty Million Theories of Grammar*. He was joking, of course, but the point was well made. Many began to question whether there would ever be a coherent theory of grammar for any language, let alone a univer-

< 21 >

sal grammar.[38] One might say that the Learning Empire has struck back, as learning theories have grown much more sophisticated, in part through the use of computer models to simulate systems with some of the complexity of human learning. No longer is it enough to base theories of learning complex things like language on rats pressing bars or pigeons pecking keys—the foundations of Skinnerian behaviorism.

This is not to say that genetics plays no role. We have created a vast linguistic environment where language permeates almost everything we do. This creates its own selection pressure, perhaps even driving the increase in brain size that so distinguishes us from the (other) great apes, whose brains are only about a third the size of ours. In technologically advanced societies, reading is critical, if not to survival then at least to effective living. Literacy emerged long after language itself and is still not universal, yet variations in reading ability seem to be genetically linked.[39] In short, we have ourselves created environments that we must adapt to, and the linguistic environment is one example. Natural selection becomes a spiral—it creates new forms of nature itself and added selection pressure.

Language is not the only product of the human ability to create complex structures. Our capacity to make machines is vastly beyond that of any other species. It is true that spiders make intricate webs, and beavers construct dams, and chimpanzees and crows fashion simple tools, but these pale in comparison with the runaway complexity of our cities, transportation devices, computers, electronic communication—and even the children's toys that I trip over after my granddaughter has invaded. Mechanical construction also builds on the past, both as individual experience and as accumulated knowledge across generations. And of course language helps too, because it enables us to share expertise and pass it on through the generations. As Isaac Newton said, "If I have seen further it is by standing on the shoulders of giants."[40]

But however we came to possess it, language does seem to have a miraculous quality, separating us from other animals and filling our lives with seemingly endless possibilities. As theorists continue

< 22 >

to grapple with its nature, language may seem to have an almost un-reachable quality, as though beyond the grasp of science or intellect. As we shall see in the next chapter, this is probably why language has long been regarded as a miracle, perhaps bequeathed by God.

The Rubicon that is language is indeed a daunting river to try to cross—even a dangerous one, as I may well discover. But that is the challenge I have set myself in this book—to explain how language might have come about through the incremental processes of Dar-winian evolution, and not as some sudden gift that placed us beyond the reach of biological principles.

< 23 >

2

LANGUAGE AS MIRACLE

In the beginning was the Word, and the Word was with God, and the
Word was God.

John 1:1

The Old Testament tells that the God gave Adam the gift of language, but then, as a self-proclaimed "jealous God," he grew
fearful that people would become arrogant and threaten his su-
premacy. So when the people began to build a tower that would reach
to heaven, he decided to take action:

> And the LORD came down to see the city and the tower, which the
> sons of men had built. And the LORD said, "Behold, they are one
> people, and they have all one language; and this is only the begin-
> ning of what they will do; and nothing that they propose to do will
> now be impossible for them. Come, let us go down, and there confuse
> their language, that they may not understand one another's speech."
> So the LORD scattered them abroad from there over the face of all
> the earth, and they left off building the city. Therefore its name was
> called Babel, because there the LORD confused the language of all the
> earth; and from there the LORD scattered them abroad over the face
> of all the earth. (Genesis 11:1–9)

Theologians have debated what that original language might have
been. The Jewish people adopted the view that Hebrew was the holy
tongue, but it was so holy that it could not be used in everyday life

< 24 >

and so died out as a spoken language for nearly two thousand years. It was revived in the late nineteenth century by Eliezer Ben-Yehuda in Palestine and is now the official language of Israel. However, when the archangel Gabriel spoke to the prophet Muhammad, he evidently spoke in Arabic, which became the accepted language of Muslims. Pious Muslims everywhere must know Arabic to read the Qur'an. In 2009 an Afghan court sentenced a man to twenty years in prison because he translated the Qur'an into the local dialect of Persian without the Arabic text alongside. Language everywhere, from Quebec to the Basque country, from Ireland to the Tamils of Sri Lanka, seems to define nationhood, and perhaps proximity to a God or Creator. A common language, though, might restore people to the state Adam enjoyed before the Tower of Babel, and—who knows?—it might even restore peace to the world.

In the past, at least, philosophers seem to have accepted that language must have been gifted to us by God. In book 3 of *An Essay concerning Human Understanding*, the English philosopher John Locke (1632–1704) assented to the divine origin but also hinted at a naturalistic cause by recognizing that language has a societal role:

> God, having designed man for a sociable creature, made him not only with an inclination, and under a necessity to have fellowship with those of his own kind, but furnished him also with language, which was to be the great instrument and common tie of society. Man, therefore, had by nature his organs so fashioned, as to be fit to frame articulate sounds, which we call words. But this was not enough to produce language; for parrots, and several other birds, will be taught to make articulate sounds distinct enough, which yet by no means are capable of language.

He went on to consider the complexities of language that render it different from the mere formation of words and therefore less parrotlike.

His earlier compatriot Thomas Hobbes (1588–1679) also agreed that "the first author of Speech was God himself" but then wor-

< 25 >

ried that God could scarcely have provided Adam with words not yet coined. He had in mind especially the words that philosophers themselves had come to use: "For I do not find any thing in the Scripture, out of which, directly or by consequence can be gathered, that *Adam* was taught the names of all Figures, Sounds, fancies, Colours, Relations; much less the names of Words and Speech, as *Generall*, *Speciall*, *Affirmative*, *Negative*, *Interrogative*, *Optative*, *Infinitive*, all which are usefull; and least of all, of *Entity*, *Intentionality*, *Quiddity*, and other significant words of the School." But I wonder if Hobbes was not quite respectful enough. If God were sufficiently powerful to have created the universe, it cannot have been too much of an impediment for him to foresee all the words that would have been necessary, with the possible exception of *twitter*.

René Descartes (1596–1650), famous for proving his own existence with the saying "I think, therefore I am," seems to have felt obliged to prove that God existed too—although there is a suspicion that this was as much out of subservience to the Church as out of logic. He was intrigued by mechanical toys, which were popular at the time, and wondered whether humans could be reduced a mechanical device. Language, he thought, was the main obstacle; in a 1646 letter to the Marquess of Newcastle, he wrote that "none of our external actions can show anyone who examines them that our body is not just a self-moving machine but contains a soul with thoughts, with the exception of words."

It was the freedom with which we use words that Descartes thought to be beyond the capability of any machine and therefore showed the existence of a nonmaterial soul, and this in turn proved the hand of God. But perhaps he secretly wondered if, after all, we are just machines, and was afraid to offend the Church. One who dared to question him was the redoubtable Princess Elizabeth of Palatine (1618–80), daughter of Frederick V of Bohemia and Elizabeth Stuart. Descartes kept up a lively correspondence with Elizabeth and indeed was said by some to be in love with her—although she was reputedly an earthy woman who didn't like to dance. In one letter she challenged Descartes as follows:

< 26 >

I beseech you tell me how the soul of man (since it is but a thinking substance) can determine the spirits of the body to produce voluntary actions. For it seems every determination of movement happens upon an impulse of the thing moved, according to the manner in which it was pushed by that which moves it, or else, depending on the qualification and figures of the superficies of this latter. Contact is required for the first two conditions, and extension for the third. You entirely exclude extension from your notion of the soul, and contact seems to me incompatible with an immaterial thing.

Elizabeth was a devout Calvinist, but her God was perhaps less fearsome than Descartes' Catholic one. And I must say I'm inclined to side with her rather than the illustrious René.

Modern-Day Miraculists

What miracle could have made you the way you are?

from the 1958 musical *Gigi*

Modern-day miraculists appeal not to so much to God as to some miracle of nature itself, perhaps some change in the wiring of the brain that endowed us with an entirely new mode of thought. It is well established, of course, that bodily changes can occur through the Darwinian processes of natural selection, but such processes are considered to be small, piecemeal, and gradual. The essence of miraculous theories is that they suppose a dramatic change to have occurred in a single step. Such theories are sometimes called "big bang" or "catastrophic" theories and are a direct challenge to Darwinian theory.

The idea that the human mind could not have evolved through natural selection is sometimes attributed to Alfred Russel Wallace. Although Wallace cofounded the theory of evolution, to Darwin's dismay he could not believe that natural selection could explain the vast difference in mental capacity between humans and apes: "Natural selection could only have endowed the savage with a brain a little

< 27 >

superior to that of an ape whereas he possesses one very little inferior to that of an average member of our learned societies."[1] His use of the term *savage* may sound elitist and discriminatory, but he was probably more enlightened than most of his contemporaries in recognizing little essential difference between indigenous peoples and those of industrial Europe. The idea of a mental abyss between the human and animal mind persists in present-day thinking, with one set of authors describing the attempt to explain it in evolutionary terms as "Darwin's Mistake."[2]

The same sentiment is presented in modern clothing by the psychologist David Premack, singling out language as the Rubicon that separates us from apehood:

> Human language is an embarrassment for evolutionary theory because it is vastly more powerful than one can account for in terms of selective fitness. A semantic language with simple mapping rules, of a kind one might suppose that the chimpanzee would have, appears to confer all the advantages one normally associates with discussion of mastodon hunting or the like. For discussions of that kind, syntactic classes, structure-dependent rules, recursion and the rest, are overly powerful devices, absurdly so.[3]

This is the *argument from incredulity*: how can so complex a function as language have evolved through the incremental steps of natural selection? A similar argument was mounted in the nineteenth century for the evolution of the eye, a complex but seemingly perfect device for informing us about the physical world. Even before Darwin, theologians such as William Paley had used the example of the eye as evidence for divine origin. That argument has now been thoroughly discredited by Richard Dawkins,[4] among others, and in any event the eye is far from perfect. Indeed its very imperfections, such as the fact that the light must pass through layers of cells in the retina before reaching the light-sensitive receptors, betray its evolutionary origins. I am especially aware of the imperfections of the eye that emerge with age as I fumble for my glasses.

< 28 >

The most prominent present-day miraculist is Noam Chomsky, whom we encountered in the previous chapter. Notwithstanding the sense that universal grammar may be a chimera, and the sense that experience may play a greater role in language acquisition than Chomskian theory implies, Chomsky and many others have held to the notion that language must have arisen uniquely, and suddenly, in our species. Chomsky assures us that this momentous event occurred even after our species itself emerged, but before our forebears migrated out of Africa to eventually populate the globe. "Roughly 100,000+ years ago," he writes, "the first question [why are there languages at all?] did not arise, because there were no languages."[5] The effect of the mental tsunami that was to come was to create not only the languages that we speak or sign but an entirely new basis of thought, available only to humans. Chomsky calls this "internal language" or *I-language*. The primary function of I-language, then, is not communication but rather thinking itself. The languages we actually speak or sign are secondary to I-language and are sometimes referred to as external languages, or *E-language*.

A prominent component of I-language in Chomsky's most recent theory is unbounded Merge, which is the basis of the notion of discrete infinity. Unbounded Merge is the contrivance that allows elements to be merged into larger units, and those larger units to be merged into still larger ones, and so on, in recursive fashion.[6] These merges occur within I-language, the language of thought itself, but are manifest in the external languages we actually speak: basic sounds (phonemes) are merged into elements of meaning (morphemes), which are merged into words, which are merged into phrases, which are merged into sentences, and onward and upward into speeches, sermons, stories, books. Applied as often as necessary, this can create structures of any desired complexity, as illustrated in the verse from "The House That Jack Built" quoted in the previous chapter. Perhaps it also enabled us to build ocean liners and skyscrapers.

So it is that we lucky humans were suddenly and uniquely blessed with a way of thinking that is entirely different from that of other species, and it was really only a kind of side effect of this that also al-

< 29 >

lowed us to express our thoughts in words—a miracle indeed, of biblical proportions, as though we crossed the Rubicon in a single leap. Moreover, not only did that miracle occur at a singular point, but it must have been bestowed initially on a single individual. Chomsky again: "Within some small group from which we are all descended, a rewiring of the brain took place in some individual, call him *Prometheus*, yielding the operation of unbounded Merge, applying to concepts with intricate (and little understood) properties."[7] Unbounded Merge is also the principal mechanism of universal grammar—the common grammatical principles said to underlie all languages. Language is the house that Merge built.

Precisely what that rewiring was is uncertain. Elsewhere Chomsky writes, "Perhaps it was an automatic consequence of absolute brain size . . . or perhaps some minor chance mutation."[8] Derek Bickerton pursues the argument from brain size, suggesting that once it had acquired a vast vocabulary of words and concepts, the brain "self-organized" to create greater efficiency and wiring economy, leading to the emergence of linguistic principles: "while a linguistic principle is unlikely to give Jack more fruitful sex, better organization of linguistic material and its processing would make Jack's brain work a lot better, thereby enhancing his overall fitness."[9]

Chomsky's theory has remarkable parallels with the biblical account. I-language may be likened to the universal language that God gave to Adam, and E-languages to the babble of tongues that emerged after the destruction of the Tower of Babel. Indeed Chomsky might have been tempted to call the lucky recipient of I-language Adam rather than Prometheus—although he was perhaps prophetic, since *prometheus* is the Greek word for forethought! Linguists are as much in dispute about the nature of I-language, or universal grammar, as were theologians about the nature of the original tongue. Chomsky's theory also has an element of Platonic idealism, and indeed of heavenly perfection, in that unbounded Merge is an ideal, a "perfect" device for creating thought processes underlying the babble of actual languages that infest the globe. The mapping of I-language onto E-language is referred to as "externalization," and it is in this

< 30 >

process that the imperfections and messiness of the world's languages arise. I-language may perhaps be likened to gravity, an elegant principle that explains why objects fall to the ground when dropped.[10] But the manner in which objects actually descend depends on lots of other things; a feather fluttering to the ground is not like a stone dropping from a tower, a bird gliding to its landing place, or a ship sinking to the ocean bed. So it is with language, which takes very different forms in the different cultures of the world.

The sheer number of actual languages, the aforementioned six thousand or so, has led some to question whether all of them can be related to a common I-language or universal grammar. As I mentioned in chapter 1, it now seems that some languages, such as those of the Piraha and Iatmul, are not recursive, a fact that raises doubts about the role of unbounded Merge, although speakers of these languages can no doubt learn other languages that do involve merging of elements. But even the parts of speech, the building blocks of theories of grammar, may not be universal. There are languages without prepositions, adjectives, articles, or adverbs, and no consensus exists among linguists as to whether all languages even distinguish between nouns and verbs.

Of more concern here, though, is the proposition that human language and thought emerged in a single step, unique to our species, and that I-language could not have evolved through natural selection. Chomsky even seems to suggest that the very atoms of I-language—the concepts that make up our thoughts—are innately given and are unique to humans. At worst this seems to pose Hobbes's dilemma: Could a mutation seventy thousand or so years ago have supplied us with all the concepts we need? Could it have foreseen the emergence of the cellphone or the Internet? Or do the internal symbols exist as empty slots waiting for externalization to fill them with reference to the experienced world? However you look at it, the idea seems to make little sense.

Like Hobbes, Steven Pinker expressed doubt, not to say incredulity, as to whether we could have been blessed with "fifty thousand innate concepts," a view he attributes not to Chomsky but to Chomsky's

< 31 >

MIT colleague the philosopher Jerry Fodor. The figure of fifty thousand is derived from the number of entries in a college-level dictionary.[11] The brain, and so the mind, is after all connected to the outside world through the senses and has powerful mechanisms for learning and remembering. This is not to say that it is entirely a blank slate, and Pinker does suggest that we may have innate concepts of substance, space, time, and causality, although even these might be supposed to have been shaped by natural selection and present in at least some nonhuman species.[12] Elizabeth Spelke and Katherine Kinzler review evidence that human infants and nonhuman primates possess "core knowledge of objects, actions, number and space," which must have deep roots in primate evolution, and perhaps earlier, but that's a long way from explaining the vast treasure-house of words that we all possess.

In support of the idea that language is inborn, a story is told of the young Thomas Babbington Macaulay, later Lord Macaulay, who at the age of four was dining with his parents at the house of a neighbor. The servant spilled a cup of coffee on his legs and anxiously inquired if he was hurt.[13] The lad immediately replied, "Thank you, Madam, the agony has quite abated." Legend has it that these were the first words that he ever spoke. While this might suggest that we inherit language fully formed, Macaulay was later infamous for suggesting that the native languages of India were inferior to those of Europe. In any case the legend is almost certainly false. It is told by a character in Randall Jarrell's 1954 novel *Pictures from an Institution* but probably goes back further than that. Another story has it that Macaulay's first words, uttered at the age of three, were "What ails wee Jock?"

Chomsky argues that language cannot have evolved through natural selection because the internal symbols of I-language have no reference to the external world and so could not have been "selected." In one recent chapter he writes: "Crucially, even the simplest words and concepts of human language lack the relation to mind-independent entities that appears to be characteristic of animal communication."[14] These symbols, he supposes, are arbitrary, with nothing in their shape

< 32 >

or sound to link them to anything in the natural world. They therefore cannot have been shaped by experience.

But the attaching of abstract sounds, such as English words, to physical entities is by no means restricted to humans.[15] The bonobo Kanzi, born in Atlanta and raised at the Yerkes Field Station in Decatur, Georgia, can understand and respond appropriately to simple spoken requests, such as "Could you carry the television outdoors?" or "Could you put the pine needles in the refrigerator?"[16] Perhaps the nonhuman word champs, though, are dogs. A border collie called Rico knows the names of two hundred objects and can fetch them on command,[17] but he is now surpassed by another border collie called Chaser, who apparently knows over a thousand spoken words.[18] These examples are probably not exceptions. They add to growing evidence suggesting that primates, at least, are capable of referring to objects in their absence, a phenomenon known as *displacement*.[19] The ability to think of objects or events in their absence is also critical to what has been termed mental time travel, and I argue in chapter 6 that mental time travel itself is not unique to humans and may indeed go far back in evolution.

Of course Chomsky's argument that internal symbols cannot have evolved through natural selection applies strictly to the symbols of I-language, not those of the external languages actually spoken or signed. But therein lies the rub. The symbols of language may have an arbitrary quality, but it need not follow that the internal representations they are associated with are also arbitrary. Chimps and dogs must surely have internal representations of the objects they attach words to, especially if the objects in question are not immediately present, and there is no good reason to suppose that these internal representations are fundamentally different from those formed in the human mind. And there is no reason to suppose that those representations emerged independently of experience. Indeed the reverse is surely true—it is precisely through interaction with the world that we and other species form representations of what is out there.

At that level at least, there is doubt as to whether there is truly a

< 33 >

sharp discontinuity between the human and animal mind, let alone the idea that language emerged in a single "big bang" less than one hundred thousand years ago.

The Archaeological Evidence

Chomsky finds additional support for the big bang theory, though, from archaeology. It is fairly generally accepted that human culture did indeed make a "great leap forward," in Chomsky's phrase, within the past hundred thousand years. This period is marked by dramatic increases in the sophistication of tools, bodily ornamentation, cave art, statuettes, burial rites, and even music. For a time it was thought that these developments, heralded as the beginnings of "behavioral modernity," were confined to Europe—a view that may echo the Victorian notion that Europeans constituted the highest form of humanity. Evidence now suggests that the great leap forward, if it is indeed a reality, can be traced to Africa, before the human dispersal around sixty thousand years ago that led eventually to human presence around the globe.[20] At least some of the markings of modernity are evident not only in Africa itself but also in other endpoints of the original dispersal, such as Australia and New Guinea.

One telling point in the big bang theory is that our other hominin cousins became extinct after the exodus from Africa, not so long after our human forebears arrived in their patch. These include the Neandertals, with whom we share a common ancestry dating back perhaps four hundred thousand years.[21] The common ancestors lived in Africa, but those who evolved to become Neandertals evidently moved out of Africa well before the human exodus and lived in Europe and Russia. Their brains were if anything slightly larger than ours, and they also made tools and hunted with spears. They managed to coexist in Europe with our own human forebears for at least ten thousand years before dying out around thirty thousand years ago—or perhaps as long ago as forty thousand years ago, according

< 34 >

to a recent estimate.[22] But that estimate may be threatened in turn by evidence that early humans mated with Neandertals. Indeed, this probably includes your own ancestors. Neandertals contributed some 1 to 3 percent to the genomes of people living outside of sub-Saharan Africa today, [23] and DNA extracted from a forty-thousand-year-old human fossil from Romania shows the fraction in that promiscuous individual to be between 6 and 9 percent.[24] Even so, whatever the furtive couplings that might have taken place, perhaps in the caves of southern France, the Neandertals were eventually unable to withstand the human invasion. In terms of the 1970s pacifist slogan "Make love not war," it may have been war that won, as it were. Or perhaps the critical advantage was more benign, a matter of wisdom rather than weaponry.

The finger points to another close relative who may have fallen victim to the wanderings of our peripatetic forebears. The discovery of a fossilized little finger in the Denisova Cave in southern Siberia seems to have belonged a young girl, sharing a common ancestry with Neandertals, but she was neither Neandertal nor human and dated from some thirty to fifty thousand years ago. She was a members of another large-brained species provisionally known as Denisovans, and DNA extracted from the fossilized finger shows that the Denisovans, too, occasionally mated with a branch of *Homo sapiens*, presumably also after the African exodus and after they subsequently arrived in Asia. Melanesians, who now inhabit Papua New Guinea and islands northeast of Australia, are said to inherit as much as a twentieth of their DNA from a Denisovan source.[25]

Of course *Homo sapiens* may have prevailed over the hapless Neandertals and Denisovans because they were somehow better adapted physically to local conditions, although this seems a little implausible given that humans were the more recent immigrants. Moreover, a proclivity for genocide is one of the less amiable characteristics of our species. But the more general answer suggested by Chomsky and others is that it was our uniquely different way to think and communicate in symbols that gave us the edge—as though we

< 35 >

talked and thought the Neandertals and Denisovans into oblivion. The archaeologist Richard Klein argues that the great leap forward may have been as late as fifty thousand years ago, and he is one of those who continue to hold that it was initially confined to Europe. He writes that it is "at least plausible to tie the basic behavioral shift at 50 ka to a fortuitous mutation that created the fully modern brain."[26] Another archaeologist, John Hoffecker, writes similarly: "Language is a plausible source for sudden and dramatic change in the archaeological record [after 40 ka] because: (a) it is difficult to conceive of how the system for generating sentences (i.e., syntax) could have evolved gradually, and (b) it must have had far-reaching effects on all aspects of behavior by creating the collective brain."[27] Again, this suggests that natural selection, the tinkerer rather than the miracle worker, could not have been responsible.

As Chomsky argued, though, if there is any truth to the big bang theory it makes better sense to place it in Africa prior to the dispersal of some sixty thousand years ago. Yet another archaeologist, Sir Paul Mellars, writes: "How far this 'symbolic explosion' associated with the origins and dispersal of our species reflects a major, mutation-driven reorganization in the cognitive capacities of the human brain—perhaps associated with a similar leap forward in the complexity of language—remains a fascinating and contentious issue."[28]

Indeed there seems to be something of a consensus among archaeologists and anthropologists that language was something a miracle. Ian Tattersall, a prominent paleoanthropologist, suggests that the emergence of language was not merely miraculous but also without precedent:

> Our ancestors made an almost unimaginable transition from a non-symbolic, nonlinguistic way of processing information and communicating information about the world to the symbolic and linguistic condition we enjoy today. It is a qualitative leap in cognitive state unparalleled in history. Indeed, as I've said, the only reason we have for believing that such a leap could ever have been made, is that it

< 36 >

was made. And it seems to have been made well *after* the acquisition by our species of its distinctive modern form.[29]

One may even suppose that the great leap forward was the very event that created us as a species distinct from those other large-brained hominins, the Neandertals and Denisovans. The anthropologist Terrence Deacon refers to us humans as "the symbolic species."[30] The British psychiatrist Timothy J. Crow suggests that the speciation event occurred about 160,000 years ago, heralding the arrival on the planet of *Homo sapiens*. He even locates the genetic source of this event in homologous regions of the X and Y chromosomes, but the evidence for this is very indirect—and indeed does smack of the miraculous. In his view, the event gave rise not only to language but also to the ability to understand what is going on in the minds of others ("theory of mind"), the predominance of right-handedness and left-brained control of language, and the human disposition to psychosis.[31] If it could do all that, it must have been a busy little gene.

Given that many humans out of Africa carry some Neandertal genes, and some others also carry Denisovan genes, we might of course question whether we are a different species at all. By one definition of the term *species*, individuals are said to belong to the same species if they are capable of successfully interbreeding, which means that we should accept the Neandertals and Denisovans into the human family. I pursue this theme in chapter 8, where I consider in more detail whether the Neandertals and Denisovans would have been capable of articulate speech. But the more important point here is that evolution is better characterized in terms of variation than of the rather arbitrary lumping of individuals into species. By this argument, the notion that humans somehow gained sudden ascendancy through a genetic big bang seems unlikely.

In any event, not all are agreed that there truly was a big bang, even in archaeological terms. The complexity of human manufacture and culture seems to increase exponentially over time, which can create the illusion that it began with a bang. Many of us may be tempted to believe there was a big bang a decade or so ago, when

< 37 >

our lives began to be ruled by computers, smartphones, Twitter, and Skype. Previous generations marveled at the telephone, the radio, or the automobile—or, let's face it, the atom bomb, which indeed created a bigger bang than had hitherto seemed possible. No doubt there are small explosions from time to time, but the overall trend is probably fairly smooth—and accelerating. Two anthropologists, Sally McBrearty and Alison Brooks, write of "the revolution that wasn't" and suggest a fairly continuous advance of technologies and cultural invention over at least the past three hundred thousand years.[32] A case in point are three well-preserved wooden spears found in a coal mine near Hanover in Germany, dating from some 380,000 to 400,000 years ago.[33] They were beautifully balanced and crafted for throwing and are associated with the fossil remains of horses, suggesting an advanced hunting technology. The hominins who inhabited this site were Neandertals (or their predecessors), not *Homo sapiens*.

The anthropologist John Shea has recently argued that the spurt of technology supposedly evident in Africa in the Middle Paleolithic was more likely a phase that came and went, an indication of variability rather than a sudden and universal shift toward greater sophistication. He goes so far as to declare that "behavioral modernity and allied concepts have no further value to human origins research."[34] We are indeed extraordinarily variable in our technologies and habits; and even in recent times the colonizers of the Americas and the European explorers in Africa found little if anything to resemble the burst of symbolic behavior in the Upper Paleolithic.

Of course all of the arguments for a big bang may well smack of the desire to assert human superiority, as though to alleviate our guilt over the disdainful way we treat other species. After all, we hunt them, eat them, ridicule them, ride on their backs, use their skins for clothing, and breed them to our specifications—and even one of our closest living nonhuman relative, the chimpanzee, is "bush meat" in parts of Africa. The issue goes deeper than whether we differ so profoundly from Neandertal or Denisovan, or the only slightly more distant *Homo erectus*. Looking not so very much further into the past on an evolutionary scale, are we really so different from our

< 38 >

close cousins the chimpanzee and bonobo, from whom we diverged around seven million years ago?

Let's see then if there are more gradual accounts of how the human mind evolved, consistent with evolution by natural selection rather than the outcome of a miracle.

< 39 >

3

LANGUAGE AND NATURAL SELECTION

Nothing in biology makes sense except in the light of evolution.

Theodosius Dobzhanski[1]

In his book *On the Origin of Species*, published in 1859 and itself something of a big bang, Charles Darwin wrote: "If it could be demonstrated that any complex organ existed, which could not possibly have been formed by numerous, successive, slight modifications, my theory would absolutely break down. But I can find no such case."[2]

If this is so, the miraculist theories of Chomsky and others, described in the previous chapter, should sound the death knell for Darwinian theory. Chomsky has repeatedly referred to language as an organ; for instance, in 2000 he wrote, "The faculty of language can reasonably be regarded as a 'language organ' in the sense in which scientists speak of the visual system, or circulatory system, as organs of the body."[3] Darwin's challenge therefore seems to apply directly to Chomsky's claims. Could this be the end of Darwin's theory?

The challenge posed by language was evident well before the modern-day miraculists showed their wares, as illustrated by the outburst from Friedrich Max Müller quoted at the beginning of chapter 1. Such was the furor that as early as 1866, just seven years after Darwin's book, the Linguistic Society of Paris banned all discussion of the origins of language, and shortly afterward the Philological Society of London of London followed suit. One might see these bans as attempts to maintain the sanctity of the human soul,

< 40 >

although they might equally have been designed to prevent discussion from degenerating into shouting matches.

But Darwin did accept that language posed something of a problem. In his 1871 book *The Descent of Man*, he wrote that "language has justly been considered one of the chief distinctions between man and the lower animals."[4] He was also clear that human language is quite unlike animal communication, despite the fact that parrots can do a passable imitation of human speech. He thought language depended on humans' "almost infinitely larger power" of associating sounds with ideas. He noted that language cannot be considered a true instinct, because every language has to be learned, but he thought that humans do have an instinctive tendency to learn language, regardless of which one they are exposed to. That, he said, was evident in the babbling of babies, and we now know too that profoundly deaf infants "babble" with their hands—presumably a precursor to sign language.[5] He concluded that language, despite its seemingly unique properties, could be explained in terms of evolutionary theory: "From these few and imperfect remarks, I conclude that the extremely complex and regular construction of many barbarous languages is no proof that they owe their origin to a special act of creation. Nor, as we have seen, does the faculty of articulate speech in itself offer any insuperable objection to the belief that man has been developed from some lower form."[6]

Nevertheless the ban on discussion of language evolution seemed to persist for much of that century—and beyond. The relative silence may also have been prolonged by Chomsky, who dominated theories of language from the late 1950s. His writing always emphasized the uniqueness of human language, with the implication that language did not evolve through natural selection, although he seemed not to make this explicit until 1975, when he wrote, "It would be a serious error to suppose that all properties, or the interesting structures that evolved, can be 'explained' in terms of natural selection."[7] His more specific ideas about the big bang that created language within the past one hundred years, chronicled in the previous chapter, seem to have emerged more recently still.

< 41 >

Spandrels and Other Potions

Chomsky's ideas were also developed at a time when natural selection itself was under attack. Stephen Jay Gould, probably the best-known evolutionary biologist of the time, argued that natural selection was but one of the forces governing evolutionary change and many adaptive changes come about indirectly. One such process is what Gould and his colleague Elizabeth S. Vrba called *exaptation*, where a structure that evolved for one use is put to a completely different use. The classic example is feathers, which originally evolved to regulate temperature but were later adapted, or exapted, for flight. Actually Darwin was aware of this phenomenon but called it *preadaptation*—a term that may have carried too much of a suggestion that evolution is somehow preordained.

A related phenomenon is what Gould and Richard C. Lewontin called a *spandrel*, a term borrowed from architecture. A spandrel is the space between two arches, or the space between a curved arch and a rectangular structure; it has no structural significance in itself, but artists have seized on the idea of creating works of art in these spaces. Having noted this in the spandrels of the Basilica of San Marco in Venice, Gould suggested that evolution might work similarly. A frivolous example is the use of the nose and ears for holding one's spectacles in place—neither organ evolved for this useful function.

Gould argued that language itself is a spandrel and went so far as to claim that "many, if not most, universal behaviors [including language] are probably spandrels, often co-opted later in human history for important secondary functions." The mind, he argued, is essentially a complex general computer, and modern-day digital computers are capable of many operations for which they were not specifically designed:

> I don't doubt for a moment that the brain's enlargement in human evolution had an adaptive basis mediated by selection. But I would be more than mildly surprised if many of the specific things it now

< 42 >

can do are the product of direct selection "for" that particular behavior. Once you build a complex machine, it can perform many unanticipated tasks. Build a computer "for" processing monthly checks at the plant, and it can also perform factor analyses on human skeletal measures, play Rogerian analyst, and whip anyone's ass (or at least tie them perpetually) in tic-tac-toe.[8]

In this view, language was one of the many "unanticipated tasks" that emerged from the enlarged and complex brain that nature bestowed on us for other purposes. Chomsky himself once adopted a very similar view: "We know very little about what happens when 10^{10} neurons are crammed into something the size of a basketball, with further conditions imposed by the specific manner in which this system developed over time."[9]

But exaptation and spandrels don't really offer alternatives to natural selection. It is of course not surprising that natural selection should build on structures that have already emerged, whether as adaptations for other functions or as spandrels. Evolution tinkers; it doesn't create. The evolution of language was no doubt constrained by the structure and idiosyncrasies of the bodies and brains of our great ape forebears. Speech, for example, takes advantage of the apparatus used for breathing, and of the larynx, which evolved originally as a valve to prevent anything but air from entering the lungs. It has also been suggested that recursion, critical to Chomsky's notion of unbounded Merge, evolved in other contexts, such as theory of mind or mental time travel—an argument I have myself developed elsewhere[10] and will visit again in later chapters. But it still requires natural selection to bring about the adjustments, just as it needed the artist to paint on the spandrels of San Marco.

In any event, language seems too universal and too central to human life to be merely an outcome of having a complex computer inside our skulls. To be sure, there are activities that are probably the outcome of our capacity for complex thought, such as playing chess or bridge, but these are not universal activities. Even mathematics is far from universal, as much of it is understood only by a privileged few. Language is everywhere, and so essential to everyday living that our capacity for it must surely have been shaped by evolution.

Some followers of Chomsky, if not Chomsky himself, have appealed to the notion of *punctuated equilibrium* to explain the suddenness with which language evolved. The idea that evolutionary change does not occur gradually and continuously but is confined to brief periodic bursts was first proposed by Gould and his colleague Niles Eldredge.[11] Such bursts typically correspond to the emergence of new species and are followed by periods of stability. The sudden appearance of language might then be taken as one such burst, supporting Timothy Crow's idea that it was language, along with other uniquely human characteristics, that effectively heralded humans as a new and different species. Gould himself, though, was fairly clear that the emergence of species was not *that* sudden; he wrote that punctuated equilibrium is "a theory about ordinary speciation (taking tens of thousands of years) and its abrupt appearance at low scales of geological resolution, not about ecological catastrophe and sudden genetic change."[12] This is certainly not consistent with Chomsky's idea that language emerged in a single person, the promiscuous Prometheus, in a single step.

The pendulum began to swing back in favor of natural selection with the publication of a landmark article by Steven Pinker and Paul Bloom in 1989, titled "Natural Language and Natural Selection,"[13] and elaborated in Pinker's 1994 book *The Language Instinct*. Exaptation, spandrels, and punctuated equilibrium, according to Pinker and Bloom, are distractions from Darwin's great insight, which is that the only way for a complex system to evolve is through a series of small mutations. Pinker and Bloom don't specify the steps through which language might have evolved, but Ray Jackendoff later set out an account that is more compatible with evolutionary theory.[14] Language is undeniably complex, so much so that linguists are still unclear as to exactly how it works. Pinker and Bloom argue that the evolution of language was driven by the extra gains from learning about the world secondhand. This is especially so in a dangerous world where

< 44 >

knowledge is often hard won, so that one's very survival may depend on sharing it. We tell our kids not to play in traffic or swim where the tidal rips are bad. Sharing knowledge is what education is all about. The sheer amount of knowledge accumulated in society is more than any one individual can acquire, so it becomes a necessity to divide much of it among individuals. So it is that we consult doctors, lawyers, engineers, accountants, architects, and even professors—not to mention soothsayers and palm readers—when we need specialized information, although the Internet may be taking over these roles.

Pinker and Bloom go on to show how the properties of language have been shaped through evolution to allow for efficient transmission of knowledge. Words, whether spoken or signed, are designed to refer to things we want to communicate about—objects, states, actions, qualities, emotions, space, and time. This is why we have different classes of words—nouns, verbs, adjectives, prepositions. These words are then combined phrases to convey propositions—things that happen, states of the world, opinions, and so forth. But even authors such as Pinker and Bloom, and Ray Jackendoff, probably underestimate the sheer variety among languages; it is a safe bet that individual languages are tailored to the individual cultural and physical needs of the peoples who use them. There may well be a common core, as implied by Chomsky's universal grammar, but so far we still have no clear idea as to how it works or how it can be understood to apply to all human languages.

To Chomsky, moreover, universal grammar is essentially "perfect," a Platonic ideal, but this is in itself counter to evolutionary theory. Pinker and Bloom cite the American evolutionary biologist George C. Williams to the effect that natural selection does not aspire to perfection.[15] Language, and the thought processes from which it derives, are sometimes kinda vague. Language thrives on variation. And so does evolution.

Evo-Devo

More recently, Chomsky has seized on another development in evolutionary theory, known as *evo-devo*, to explain the apparently sudden emergence of language. Evo-devo is effectively a synthesis of evolutionary principles with those involved in the development of the individual during the lifespan. This emerged from the discovery of Homeobox (Hox) genes, present in virtually all species, which regulate expression of other genes, such as those involved in the development of the eyes, limbs, or heart. In this view new structures can emerge through the changes in the timing or sequence with which other genes are switched on or off, rather than from the emergence of new genes to create new functions. It has been widely assumed that we humans are so complex that we must have many more genes than other species do, but studies of the human genome have suggested that we are endowed with many fewer genes than expected. The exact number is still in dispute—perhaps only around twenty thousand—but it appears that the number is little more than in the roundworm. That we differ from roundworms must be due largely to the influence of regulatory genes, although we should perhaps not underestimate the rich inner life of the otherwise rather passive worm.[16]

According to Chomsky, evo-devo implies that "slight changes in the hierarchy and timing of regulatory mechanisms might yield great superficial differences—a butterfly or an elephant, and so on."[17] A similar slight change in regulatory processes might then have given us the miracle of universal grammar.

But butterflies weren't turned into elephants, or even princesses into frogs, in a single step. Morten Christiansen and Nick Chater explain that evo-devo can't explain a sudden appearance of anything as complex as universal grammar (UG):

> Small genetic changes lead to modifications of existing complex systems, and these modifications can be quite far-reaching; however, they do not lead to the construction of new complexity. A mutation might lead to an insect having an extra pair of legs, and a com-

< 46 >

plex set of genetic modifications (almost certainly over strong and continuous selectional pressure) may modify a leg into a flipper, but no single gene creates an entirely new means of locomotion, from scratch. The whole burden of the classic arguments for UG is that UG is both highly organized and complex, and utterly distinct from general cognitive principles. Thus, the emergence of a putative UG requires the construction of a new complex system, and the argument sketched above notes that the probability of even modest new complexity arising by chance is astronomically low.[18]

Pinker and Bloom, champions of Darwinian theory, are nevertheless adamant that even our closest nonhuman relatives, the chimpanzee and bonobo, do not have language. The question therefore remains as to how language could have evolved in our species but not in any other. But as Pinker and Bloom point out, there is time between the separation of hominins from the other great apes for language to have come about through natural selection rather than exploding in a big bang. They write: "As far as we know, this would still leave plenty of time for language to have evolved: 3.5 to 5 million years, if early Australopithecines were the first talkers, or, as an absolute minimum, several hundred thousand years in the unlikely event that early *Homo sapiens* was the first."[19] From what we know now, we could even push it back a bit further to something like six or seven million years; the "several hundred thousand" is probably not more than two hundred thousand, but even this rules out the idea that language emerged within the last hundred thousand years.

Of course, if language is so adaptive, one might ask why it has not emerged in any other species. One might perhaps grope toward an answer in terms of "special" ecological circumstances that our hominin forebears found themselves in, although there is inevitably a sense of special pleading—a Just-So story. Perhaps with the advent of a change from a forested environment to one of more open country from perhaps six million years ago, our hominin forebears were more exposed to the dangerous predators—lions, leopards, hyenas, or their prehistoric predecessors—that roamed the African savanna.

< 47 >

Another view is that our forebears inhabited coastal areas rather than the savanna, foraging in water for shellfish and waterborne plants, and that this had a profound influence on our bodily characteristics and even our brain size.[20] This aquatic phase may perhaps have preceded a later transition to the savanna—a possibility discussed further in chapter 5. Either way, the change from a forested environment required new adaptations, including the emergence of what has been called the "cognitive niche,"[21] with survival depending on social cohesion. More effective communication would have been critical. This development probably evolved from around 2.9 million years ago, during the era known as the Pleistocene, when brain size increased dramatically.

It is also possible that language is not really so adaptive after all and may lead to the destruction of our species, but I leave that dark thought to the final chapter.

In any case, even Pinker and Bloom may have insisted too strongly that language emerged entirely new in our own genus. They reject the well-known studies of Sue Savage-Rumbaugh and others on the acquisition of forms of visible language, some of it based on sign language, in other great apes.[22] While it is true that these studies demonstrate little grammatical ability, or even any ability to converse, the biological roots of language may well go back far in primate evolution. In two recent books, James Hurford has argued that the origins of language can be traced back to primate communications, and that the evolution of language is at least potentially explicable in terms of natural selection.[23]

Another author to adopt an evolutionary approach is Tecumseh Fitch in his 2011 book *The Evolution of Language*,[24] which nevertheless leaves room for the "catastrophic" development implied by Chomsky's theory. He distinguishes between the *faculty of language in the broad sense* (FLB), which covers the external aspects of languages and their overlap with animal communication, and the *faculty of language in the narrow sense* (FLN), which incorporates the aspect of language that is unique to humans.[25] In essence, FLN is Chomskian I-language, also identified as universal grammar, or in the minimal-

< 48 >

ist theory as unbounded Merge. While Fitch's book is mostly about FLB, he does retain space for FLN and the question whether it really did emerge catastrophically in human evolution. At one point he acknowledges that the apparent complexities of syntax may not have arisen uniquely in our species after all but may derive from "much older mechanisms" that were already present "before human language evolution began."[26] To the extent that this might be so, it is not out of line with the theme developed in this book.

The Diversity of Languages

As we saw in chapter 1, some six thousand languages coexist in the present-day world. Their widely differing characteristics pose severe problems for the notion of universal grammar, to the point that some authors declared that notion to be dead—or to have no clothes. By the same token, the vast differences make it difficult to accept that language evolved in a single step, even in that single individual whom Chomsky whimsically named Prometheus.

One possibility, though, is that languages have mutated from a common source. Biological mutation has created some eight million species in the world, all perhaps derived from a single event that created a self-replicating cell. Through mutations and natural selection, biological forms multiplied and diversified. And there is of course an astonishing diversity of organisms, ranging from amoebas to apes, bananas to beeches, herrings to humans. Could the diversity of languages have arisen in the same way but through cultural rather than biological change? Languages do appear to mutate over time, with different forms selected according to the requirements of different groups. By examining similarities between different languages, one can attempt to trace their common ancestries, with the hope of eventually finding that original source whence all languages evolved.

Attempts to do this have been controversial; indeed, in banning discussion of language origins in 1866, the Linguistic Society of Paris also banned any claims as to the existence of an original language, a

< 49 >

universal mother tongue. Some have nevertheless claimed that it is possible to date the earliest language at some 50,000 years,[27] although a more accepted view is that studies based on patterns of similarity among present-day languages cannot take us nearly so far back. Jared Diamond, for instance, writes of the so-called "5,000–10,000-year barrier" but gives an example which may take it back to perhaps 12,000 years.[28] Of course this cannot be the date of the supposed original tongue, because humans had dispersed around the globe well before that—it is rather a reflection of the date at which the signal from present-day languages is effectively lost. But in principle, at least, it is possible that there was an original language, which is sometimes spoken of as ProtoWorld. And those who have tried to trace the mother tongue have neglected signed languages, an omission rectified later, in chapter 7.

One way in which spoken languages vary is in the basic sounds that make up spoken words. Nicholas Evans asserts that there are well over fifteen hundred possible speech sounds, but no language uses more than 10 percent of them for its inventory of phonemes.[29] The most parsimonious of spoken languages appears to be that of the women of the Pirahã, a small hunter-gatherer Amazonian tribe in Brazil.[30] The Pirahã language as spoken by women has only seven consonants and three vowels. As though to compensate for a male lack of verbal fluency, the men are permitted eight consonants and three vowels; with eleven phonemes, male Pirahãns tie with Hawaiians for the second most parsimonious language. Actually, there may be an even simpler language. Silbo Gomero is a whistled language used by shepherds on the island of Gomero in the Canary Islands, which is reduced to two vowels and four consonants. It is perhaps an unfair example, because it is based on Spanish and is essentially a cutdown version of it. Brain imaging shows, though, that when people who speak the language listen to it, the brain areas responsible for comprehending spoken languages are activated. The brain, at least, thinks it's a language—and who or what to know better?[31]

In any event, the sophistication of a language probably doesn't depend much on the number of phonemes employed. New Zea-

< 50 >

land Maori has only fourteen phonemes, but Maori are known for fine oratory. In Maori society, as in other traditionally oral societies, speech implies power and status; the New Zealand scholar Anne Salmond writes that among the Maori "oratory is the prime qualification for entry into the power game."[32] The most phonologically diverse language may be !Xóõ, which is spoken by about four thousand people in Botswana and Namibia and has somewhere between 84 and 159 consonants. English bumbles along with about 44 phonemes.

This variation can provide clues as to when spoken language began. The number of phonemes tends to be smaller in small populations, suggesting that as small groups migrated away from parent populations, the number of phonemes they used tended to diminish. Quentin Atkinson observes that the number of phonemes in different languages tends to diminish with distance from Africa, suggesting that spoken language originated in Africa and spread from there as humans migrated to other parts of the world. Working backwards suggests an African founder effect for languages roughly in conformity with the "big bang" theory discussed in the previous chapter. Atkinson concludes as follows: "An origin of modern languages predating the African exodus 50,000 to 70,000 years ago puts complex language alongside the earliest archaeological evidence of symbolic culture in Africa 80,000 to 160,000 years ago. Truly modern language, akin to languages spoken today, may thus have been the key cultural innovation that allowed the emergence of these and other hallmarks of behavioral modernity and ultimately led to our colonization of the globe."[33]

But the mutation story may be too simple. Other evidence suggests that the diversity of languages need not reflect a progression from one to many. For most of human history, people lived in small tribes or bands, partly in isolation but also often in conflict with one another. This meant that means of communication probably evolved within these small communities and developed their own idiosyncratic properties. There may even have been pressure to develop forms of language incomprehensible to rivals, a kind of secret code. Even today, great diversity can exist within small geographic areas.

< 51 >

In chapter 1 I noted the extreme example of Vanuatu, with an area of only about 4,379 square miles, which is host to over one hundred different languages.[34] Moreover, we also saw in chapter 1 that there is an inverse relation between cultural sophistication and linguistic complexity: the simpler the culture in material terms, the more complex the language, especially at the level of individual words. Thus the languages of simpler cultures tend to pack grammatical information into single words, whereas those of industrial society tend to use separate words in combination to create grammatical distinctions. I gave the example *Ęskakhǫna'tàyęthwahs* from the Cayuga of North America, which in English can be translated as "I will plant potatoes for them again."

Rather than emerging as a single entity, then, language may well have arisen within small groups, perhaps even within families, and developed idiosyncratic vocabularies and grammars quite early on. Later on as groups amalgamated, or as dominant groups colonized small ones, languages decreased in number and simplified. A powerful but recent influence may have been what we are pleased to call civilization. As people began to mingle with others who spoke different languages, the common language that they came to share became simpler. Even within the past two thousand years or so, languages have lost much of their complexity, at least at the level of words (morphology). The classical languages like ancient Greek and Latin had complex systems of inflections, with many different forms of nouns and verbs to indicate their roles in sentences. Nouns were inflected to indicate differences between subject and object, number, and so forth, and verbs to indicate tense and mood as well as other more complex roles. Much of this is stripped away in modern-day Greek, Italian, or French. Modern English was forged from contact between Norman French and Anglo-Saxon and is morphologically simpler than either. Some have noted an even simpler form of English, known as Chinglish, which seems to be emerging from the Chinese influence in the English-speaking world. Compared standard English, Chinglish pretty simple.

It is therefore possible that language emerged in Africa, not from a

< 52 >

single source, let alone a single Prometheus, but framed within different groups and shaped to meet the various communication and cultural needs of those groups. There were probably mutations and development of individual languages over time, but also simplification and mixing. Discussing the manner in which languages mold themselves to cultural influences, Morten Christiansen and Nick Chater effectively reverse the Chomskian view that a dramatic change in the brain shaped language, arguing instead that language itself adapted to the constraints of the brain. This view of language itself as an "organism," to use Christiansen and Chater's term, seems to allow the possibility that language evolved gradually and not as a "big bang," with precursors to be found in our common ancestry with other primates—and perhaps even further back in our heritage.

Although Chomsky has often declared language to be an organ, comparable to the heart or liver, Christiansen and Chater do not use the term *organism* in the biological sense. Rather, language is a meme, shaped by cultural instead of biological evolution. They do concede that biological capacities peculiar to humans caused language to take the various shapes it did, but these capacities were not specific to language; they included increased capacity to hold material in working memory and increased ability to simply learn words and sequences. In these respects language can be regarded as a tool, and indeed the evolution of tools in general must also have depended on biological prerequisites, such as enhanced manual ability, powers of imagery, and ability to plan. Thus the Israeli linguist Daniel Dor, in his 2015 book *The Instruction of Imagination*, refers to language as "social communication technology,"[35] while Daniel Everett, in *Language: The Cultural Tool*, writes that "languages are tools. Tools to solve the twin problems of communication and social cohesion. Tools shaped by the distinctive pressures of their cultural niches—pressures that include cultural values and history and which account in many cases for the similarities and differences between languages."[36]

These remarks also point to a change in the definition of language itself. Where Chomsky defined language as internal to the mind, a mode of thought rather than of expression, authors like Dor and

< 53 >

Everett return to the more commonsense view of language as communication. In Dor's words, language is "the sharing of experience."[37] In this view, thought itself is not language, but language is shaped to allow us to express our thoughts. Before discussing this, we need to consider the nature of thought. This is the topic of the next three chapters.

< 54 >

PART TWO

The Mental Prerequisites

One of the foundations of Chomsky's theory is that language is a mode of thinking, an internal system, rather than a communicative one. In part 2 of this book I document the internal structures on which language depends, but I argue that these are not specific to language, nor did they emerge suddenly in our own species. Rather, they are the outcome of normal evolutionary processes of mental adaptation to a spatial world with growing social imperatives. In this view language is shaped to allow us to share internal thoughts, and indeed may be uniquely equipped to do so in our species. In order to understand how it does this, we must consider the nature of thought itself. Indeed, as we shall see, some of the attributes of language are indeed dependent on the nature of our thoughts and experiences, although not in the manner proposed by Chomsky.

Chapter 4 argues for a clear distinction between language and thought, and it develops the argument that one aspect of thought that shaped language as a communication system is the ability to travel mentally in time and place. This leads to the feature of language known as *displacement*, allowing us to refer to events that are nonpresent. We can describe events that occurred in the past or that are planned for the future, or even events that are purely imaginary. I will suggest that the capacity to travel mentally in time is not unique to humans but evolved incrementally throughout the evolution of animals that inhabit and move around in spatial environments. What does seem to be distinctive to humans is not mental time travel itself but the capacity to share our travels.

< 55 >

In chapter 5 I discuss another kind of mental travel, our excursions into the minds of others. This is known as *theory of mind* and is also critical to language. Indeed, effective communication is at best crippled if speaker and audience do not share a stream of thought that is only partially determined by language. Theory of mind may well be more highly developed in humans than in other living species, although there are at least some indications that primates, and especially the great apes, have limited access to what others are thinking or feeling.

Chapter 6 brings mental time travel and theory of mind together into perhaps the most distinctive of human activities, the telling of stories. Of course stories themselves are generally told through language, although in the modern age movies, videos, and TV soaps—not to mention comic strips—can add a substantial nonlinguistic element. The basis of stories is the mental journeys we are able to take into other places, other times, and other minds, often into realms that are pure fantasy. Although language is critical to the telling of stories, it is distinct from the stories themselves.

< 56 >

4

THINKING WITHOUT LANGUAGE

We should have a great many fewer disputes in the world if words were taken for what they are, the signs of our ideas only, and not for things themselves.

John Locke[1]

Do we think in words? John Locke aside, many philosophers and linguists have argued that we do. Plato wrote that "the soul when thinking appears to me to be just talking,"[2] and the German philosopher Immanuel Kant wrote, "Thinking is speaking to ourselves."[3] The founder of behaviorism, John B. Watson, was evidently unwilling to admit mental processes at all and equated thought with subvocal speech—we think, he thoughtlessly thought, by muttering inaudibly to ourselves. And as we saw in chapter 2, even Noam Chomsky, arch-critic of behaviorism, argued that thought and language share a common basis in I-language. This is encapsulated also in the so-called language-of-thought hypothesis (LOTH) proposed by the philosopher Jerry Fodor, who even argued that all of the concepts underlying words are inborn[4]—the "fifty thousand innate concepts" ridiculed by Steven Pinker.[5]

Merlin Donald offers a different perspective, arguing that the arrival of language effectively strangled the "language of thought": "Once the mind starts to construct a verbally encoded mental 'world' of its own, the products of this operation—thoughts and words—cannot be dissociated from one another. . . . The models and their

< 57 >

words are so closely intertwined that, in the absence of words, the whole system simply shuts down. There is no surviving 'language of thought' from which the words have become disconnected, no symbols, no symbolic thought, no complex symbolic models."[6]

Words are indeed heavily intertwined with our mental concepts, which is why we are so readily able to tell others of our experiences. Words may even help refine and shape our concepts. But they are not the concepts themselves, and they can become dissociated from them, as when we can't remember a name or when we entertain ideas for which there are no words.

Some construe thinking in language as thinking in abstract symbols, as in Chomsky's I-language, rather than in images. In recent times, at least, the idea that we think in abstract symbols, whether words or some deeper entities endowed by our genetic makeup, owes something to the digital computer, in which everything is effectively encoded abstractly as strings of ones and zeros—a kind of Morse code. It was Alan Turing, one of the famous group of scientists and mathematicians who worked at Bletchley Park in England to break enemy codes during World War II, who first saw the potential of computational methods to simulate the human mind. He devised the eponymous Turing test, in which a human judge is asked to decide whether a series of responses to questions is produced by a person or by a machine. If the person cannot tell the difference, the machine is said to have passed the test.[7] It has a mind—or is at least fooling people into believing it has one.

Turing's insights led to the science of artificial intelligence. The idea that the mind was simply a complex computer took hold from about the 1950s and effectively transformed psychology from the behaviorism that had held sway, especially in the United States, from early in the century. Computers were programmed to simulate such human activities as problem solving,[8] and even visual imagination came to be regarded as fundamentally computational rather than picturelike.[9] In 1957 Herbert Simon predicted that within ten years a computer would routinely beat the world's best chess players. He was wrong; it took until 1989 for the computer Deep Thought II to

< 58 >

systematically beat the reigning Scottish champion, David Levy. Still, that's not bad—and it must be said that computers can now beat anyone. Chomsky based his linguistic theories on the notion that a computer might be programmed to determine whether a given sentence was grammatical or not—a sort of reverse Turing test. In the brave new computational world we could perhaps get rid of people altogether and have the earth populated by intelligent robots. This could solve problems of famine and overcrowding, war and pestilence, and the geeky robots might enjoy games of chess and dancing to electronic music.

Of course artificial intelligence is by no means dead, and computational procedures are doing wondrous things in replacing people. Google is a prime example—you can ask it almost anything and find an answer. But much of the power of computation rests on an enormous memory and rapid search, and rather little on understanding.

Common sense surely tells us that we can think without language or digital pyrotechnics, but rather in terms of images that bear the hallmarks of experience in the world around us. As age advances upon me, in fact, I find myself increasingly less able to remember people's names, but I have clear images of what they look like—or used to look like. As well as being excessively verbal creatures, we are highly dependent on the visual sense, and many of our thoughts seem primarily visual. Imagine, for example, your movements this morning since getting out of bed: perhaps taking a shower, fetching the paper, having a croissant and coffee before dashing to work. Oh, and feeding the cat. These thoughts are likely to be mainly visual, although you might also imagine the feel of the hot shower or the taste of the croissant, or even a shiver of distaste at what the cat is eating. But it is unlikely that words intruded to any great extent—except perhaps as you glanced through the paper. With respect to visual aspects, at least, brain imaging shows that the parts of the brain activated by such imaginings activate the same parts of the brain that are activated by vision itself, as well as the parts involved in memory, but the parts involved in language are relatively undisturbed.[10]

Ray Jackendoff also points out that thought cannot be tied to any

< 59 >

specific language, such as English or French. Rather, thought is what is supposed to remain constant when we translate from one language to another, and so it belongs to neither.[11] He goes on to suggest that the thought itself is unconscious and what we experience is its "externalization" in language. But I don't think this is entirely correct, because the thought you are trying to express is often visual, or even auditory as you mentally replay a favorite tune—or imagine your movements after getting out of bed, as in the previous paragraph. Such thoughts are conscious.

But there are also thoughts that are neither imagelike nor linguistic—what the Wurzburg school, founded by the psychologist Oswald Kulpe in 1896, called "imageless thought." I'm sure we all have trouble from time to time turning thoughts into words, as though there simply aren't the words to express them—or else we can't find them—nor is there any image to call upon. One example is the "tip-of-the-tongue" phenomenon, where we have some concept in our head but can't quite find the word that describes it. But the problem goes beyond word finding; we sometimes simply don't have the words at all. Curiously, even Chomsky has articulated this very problem, seemingly in contradiction to his notion that language and thought share the same foundation:

> Now what seems to me obvious by introspection is that I can think without language. In fact, very often, I seem to be thinking and finding it hard to articulate what I am thinking. It is a very common experience at least for me and I suppose for everybody to try to express something, to say it and to realize that is not what I meant and then to try to say it some other way and maybe come closer to what you meant; then somebody helps you out and you say it in yet another way. That is a fairly common experience and it is pretty hard to make sense of that experience without assuming that you think without language. You think and then you try to find a way to articulate what you think and sometimes you can't do it at all; you just can't explain to somebody what you think.[12]

< 60 >

As Steven Pinker points out, language itself is fairly useless as a medium for thinking or reasoning, or even understanding.[13] One obvious difficulty is polysemy, the fact that many words have different meanings—the word *present*, for example, can refer to a point in time (right now), a gift, the act of showing or displaying, or the state of being physically located somewhere. I'm told that the worst culprit in English is the word *set*, which takes up the most space in the dictionary because of its many meanings. The definition from the nicely compact *Chambers Dictionary* on my cellphone gives forty-four different meanings of *set* as a transitive verb, fifteen as an intransitive verb, twelve as an adjective, and thirty-four as a noun. In such cases, people must extract the appropriate meaning in order to understand, or use that understanding to reason. The words themselves are inadequate; it's the underlying thoughts that count. Pinker quotes from the poem "Limitation" by Siegfried Sassoon:

> Words are fools
> Who follow blindly, once they get a lead.
> But thoughts are kingfishers that haunt the pools
> Of quiet; seldom seen . . .

Of course the relation between language and thought does depend on how language itself is defined. If language is simply internal thought, as in Chomsky's I-language, then there is indeed a fundamental identity between language and at least some aspects of thought, but, as explained earlier, language in this book is regarded as a system of communication and not as a mode of thought. The definition of language as the sharing of experience, as Daniel Dor notes in his book *The Instruction of Imagination*, "turns the Chomskian proposal on its head."[14] In Chomskian terms, then the focus then shifts from I-language to E-language, the external forms of language that Chomsky considered relatively trivial and uninteresting. Even so, the structure of thought is critical to the nature of language, even if it is not language itself. But the critical shift is that thought itself is here

< 61 >

regarded as the outcome of natural selection and not as a sudden catastrophic change.

That said, language does vastly influence our thoughts. A large proportion of our knowledge is received from others, mostly through the medium of language. It is through language that we learn things from our parents, schools, and friends. We learn about people through gossip. The Russian psychologist Lev Vygotsky went so far as to declare that "all the higher functions originate as actual relations between human individuals,"[15] in which language plays the major role. Purely as a system of sharing, language is highly adaptive. As Steven Pinker and Paul Bloom put it, "Children can learn from a parent that a food is poisonous or a particular animal is dangerous; they do not have to observe or experience this by themselves,"[16] Utterly critical to human life and development are stories. The developmental psychologist Katherine Nelson writes of "storied thoughts," in which children's thinking becomes "narrativized" through storytelling.[17] Stories are of course "told," but the story itself is understood as a sequence of events, not of words, and there are many different ways to put stories into words—or indeed into film, opera, or ballet. The story of stories is covered in detail in chapter 6.

Words are entities, just as objects are, and we can of course have words in mind, just as we can have images of objects in mind. But thinking *of* words need not imply thinking *in* words. I have words in mind when preparing a talk or struggling to write sentences, as I am doing now, but I still need to have appropriate thoughts before finding words to express them—what the linguist Dan Slobin called "thinking for speaking."[18] Words and concepts dance together in the mind, but they sometimes become disconnected, as when we can't remember someone's name or the word for an elusive concept—like *schadenfreude*. (There, I got it).

Language, then, is a system that can be attached to our internal thoughts allowing us to share those thoughts with others—a notion explored in part 3 of this book. It undoubtedly has properties unique to humans, but the nature and origins of thought itself probably go far back in evolution. Of course many of the things we think about

< 62 >

are unique to our species, and unique to the particular cultures we belong to, but we are probably not alone in the ability to form concepts about things or events in the world, or even to mentally construct scenes, as I suggest below. Something of a window on the nature of thought is provided by a phenomenon that I'm sure is familiar to all.

Mind Wandering

If we are left to think undisturbed, without focusing on the immediate environment, our minds wander. These wanderings include "autobiographical memory retrieval, envisioning the future, and conceiving the perspectives of others."[19] Such activities are also the stuff of fantasy and fiction—points I have elaborated elsewhere.[20] Brain imaging shows that our mind wanderings activate a widespread network in the brain, in which the frontal and parietal lobes play a major role. This network has been dubbed the *default-mode network*.[21]

In functional magnetic resonance imaging (fMRI), activity in the brain while people perform designated tasks is measured in terms of blood flow to active regions, a technique made possible by the fact that the blood contains iron so that its movement can be detected magnetically. But when people are not engaged in a task, the flow of blood to the brain is reduced by only about 5 or 10 percent and is distributed more diffusely through default-mode network. This is when the mind is released from the task in hand and starts to wander. Mind wandering takes us away from the bustle of the present and into the nonpresent—whether in the form of remembered experiences, planned future ones, or happenings that are purely imaginary. It can take us into the minds of others, as when novelists write from the points of view of other people. Indeed, the archetypal mind wanderer is himself a fictional entity, the protagonist of James Thurber's story "The Secret Life of Walter Mitty." Here he is in full steam:

"We're going through!" The Commander's voice was like thin ice breaking. He wore his full-dress uniform, with the heavily braided

< 63 >

white cap pulled down rakishly over one cold gray eye. "We can't make it, sir. It's spoiling for a hurricane, if you ask me." "I'm not asking you, Lieutenant Berg," said the Commander. "Throw on the power lights! Rev her up to 8,500! We're going through!" The pounding of the cylinders increased: ta-pocketa-pocketa-pocketa-*pocketa-pocketa*. The Commander stared at the ice forming on the pilot window. He walked over and twisted a row of complicated dials. "Switch on No. 8 auxiliary!" he shouted. "Switch on No. 8 auxiliary!" repeated Lieutenant Berg. "Full strength in No. 3 turret!" shouted the Commander. "Full strength in No. 3 turret!" The crew, bending to their various tasks in the huge, hurtling eight-engined Navy hydroplane, looked at each other and grinned. "The old man will get us through," they said to one another. "The Old Man ain't afraid of Hell!" . . .

"Not so fast! You're driving too fast!" said Mrs. Mitty. "What are you driving so fast for?"[22]

Walter Mitty was daydreaming, but a purer form of mind wandering is dreaming while asleep, where the ambient world is all but excluded. We all dream three or four times a night. Our dreams consist of often bizarre rearrangements of past episodes or imaginings, and most of them are simply forgotten. During dreaming, the motor system is inhibited, which is fortunate because it prevents us from acting out our dreams, although this can also have the disturbing (if unreal) effect of preventing us from running away from some horrendous but imagined threat—the mares of the night.[23] The often bizarre nature of dreams is not well understood but may capture the essence of the random constructive element that underlies some of our waking thoughts as well.[24] It provides the creative component of art, music, and fiction and also pervades our attempts to understand reality. Writing about Albert Einstein, the Italian physicist Carlo Rovelli remarks, "You don't get anywhere by not 'wasting' time."[25]

Although there are constructive aspects to mind wandering, there is also something of a downside. As Walter Mitty's wandering sug-

< 64 >

gests, it can be dangerous when driving or when it distracts from some other critical task. People tend to be less happy when their minds wander away from a task at hand, perhaps because much of our wandering consists of rumination over past indiscretions—those embarrassing moments we can't seem to get out of our minds. When this happens it's perhaps a good idea to try to wander into happier territory. Some religions, such as Buddhism, use meditation to encourage "mindfulness" through meta-awareness of ongoing thoughts themselves, so the mind is prevented from wandering to other topics.[26]

Mind wandering is dependent on memory. The psychologist Endel Tulving of the University of Toronto distinguished *episodic memory*, which is memory for specific events from the past, from *semantic memory*, which is general knowledge about the world.[27] The two are not completely independent, as our memory of events is informed by our knowledge of the world—memory of a visit to a restaurant, for example, is partly sustained by what restaurants are and what happens in them. Nevertheless, episodic memories form a fair chunk of our mind wandering as we replay past events, whether in triumph over a past success or despair over a past embarrassment. Stanley Klein suggests that conscious recollection is part of the process by which we form the concept of self, at the same time arguing that there is no unified "I" to be found in our recollections.[28] We perhaps construct different senses of self from different recollections and different fantasies.

Men who served in wars may remember acts of heroism that were in fact less heroic than remembered or perhaps did not even occur at all. In his presidential campaigns, Ronald Reagan often spoke of war-time heroics, but at least some of the episodes he described actually came from old movies. The English poet John Fuller writes:

> I am your memories. They are not me.
> So it feels strange to be remembered by
> These relics of my personality.[29]

< 65 >

Because they have relatively little to do with forming a diarylike record of our past, and more to do with providing scenarios from which to build possible future episodes and perhaps with creating gratifying self-images, episodic memories are notoriously unreliable. This is the bane of courts of law, where judgments of guilt or innocence often depend on how accurately past events are remembered. "The past is a foreign country," wrote L. P. Hartley in his novel *The Go-Between*, "they do things differently there." Hillary Clinton told of landing in Bosnia in 1996 under sniper fire, but television records show her being greeted in peace by a smiling child. Elizabeth Loftus, a pioneer in the study of false memories, vividly recalls discovering her mother lying dead in a swimming pool. It turned out that the memory was false. She was in fact asleep at the time her mother's body was discovered, by her aunt.[30]

Endel Tulving has persistently claimed that episodic memory is something that only humans possess. This does not mean that other animals don't have memory. The distinction between episodic and semantic memory can be described as that between remembering and knowing. A dog that buries a bone may later know where the bone is buried but not remember actually burying it. Many birds have knowledge of multiple locations where they have cached food, but in Tulving's view they would have no memory for the individual acts of caching. The Clark's nutcracker is said to hide around thirty-three thousand seeds in around seven thousand locations every fall, and relies on its spatial memory to find them over the winter.[31] That is prodigious enough without the added burden of remembering the actual episodes of caching.

Episodic memory implies the ability to travel mentally back in time. As the Queen in Lewis Carroll's *Through the Looking Glass* remarked, though, "It's a poor sort of memory that only works backwards." Tulving later extended the idea of mental time travel to including traveling into an imagined future,[32] an idea elaborated by Thomas Suddendorf and me.[33] In this view episodic memory did not evolve simply to provide a record of the past but serves as building

< 66 >

blocks from which to construct possible futures. And it is the future, not the past, that really matters; Klein and colleagues have even suggested that "memory enabled humans to be aware of the future before they were able to consciously experience the past."[34] Regardless of which came first, then, episodic memory is part of a more general capacity for mental time travel, whether back into the past or forward into the future. At a talk I gave recently, a woman in the audience asked me how a memory of the episode of her asking might prove beneficial to me in the future. I told her that I would undoubtedly encode the incident, and if I were to meet her again I would be wary of being asked a difficult question. I would also be wary in any similar context of that particular question—it's always useful to be prepared for specific questions. I might even use the episode as a useful illustration in a book . . . And needless to say, I remember the episode well, even after several months.

Following from Tulving's view that episodic memory is unique to humans, Suddendorf and I proposed that only humans have the more general capacity for mental time travel. The idea is actually an old one. Robert Browning's 1885 poem "A Grammarian's Funeral" includes the following lines:

> He said, "What's time? Leave Now for dogs and apes!
> Man has Forever!"

The poet W. H. Auden is only slightly more generous to animals, although less so to humans, in a poem ironically titled "Progress?" and published in 1972:

> Sessile, unseeing,
> the Plant is wholly content
> with the Adjacent.
> Mobilized, sighted,
> the Beast can tell Here from There
> and Now from Not-Yet.

< 67 >

Talkative, anxious,
Man can picture the Absent
and Non-Existent.

If animals are unable to travel mentally in time, this need not mean that they cannot prepare for future events. When a bird caches food, it is effectively planning for the future occasion of recovering that food, even if it cannot bring to mind the caching itself or the future occasion. The dog that knows where the bone was buried without remembering the act of burying it is nonetheless prepared for a future act of recovery. Evolution itself is a recipe for the future. Animals have evolved instinctive behaviors on the basis that they increase the chances of future survival, but this need not mean that they envisage the future. The godwits that show up in New Zealand after the long migration from Alaska probably do not conjure up images of the pleasant environment they will find in my country when they set out. They just goof off in a southwesterly direction, guided by instinct rather than by dreams of Auckland nightlife.

The capacities to learn and remember are also oriented to future survival, but these need not imply the imagining of future events either. The hapless pigeon in the Skinnerian laboratory that learns to peck a key to receive a food reward is ensuring a future supply of food, if not of hap, but may or may not envisage the arrival of either. Many animals, including rats, learn about their spatial environments, enabling them to navigate in future. Much of the knowledge we humans have acquired is also a recipe for future action but does not require the active replaying of the source of that knowledge—knowing the sun will rise tomorrow does not depend on mentally replaying its rising yesterday. The question, then, is whether mental time travel, the actual imagining of past and future scenarios, is a faculty that is indeed uniquely human, as Suddendorf and I suggested.[35]

I suspect, in fact, that Suddendorf and I probably overstated the case and that there are at least ripples of mental time travel in other species, even in birds. And rats.

< 68 >

Mind Wandering in Animals

Nicola Clayton and her associates at the University of Cambridge have shown that scrub jays, when they cache food, seem to remember not only *where* they hid the food but also *when* they hid it. If they have recently hidden worms and nuts, they will retrieve the worms in preference to the nuts because they like them better. But if more time has elapsed they will prefer the nuts, because the worms will have decayed and become unpalatable. This needn't mean that they recall the actual episode of hiding these items of food, but it does suggest that they know more than the mere location of the food. If part of the memory consists of the time at which the food was cached, then it is at least somewhat dependent on the caching episode itself.

Clayton and colleagues have also shown that if birds are watched by another bird while they cache food, they will later recache it in another place. This suggests that they envisage the watching bird returning later to steal the food, so they had better put it somewhere else. They seem to do this, though, only if they have themselves stolen food. As with our own species, it takes a thief to know a thief. This study might suggest that these little birds have a sense of future as well as past and that their minds can even wander into the mind of another bird.[36]

But flights of fancy notwithstanding, we are not closely related to birds and should probably look to our closer relatives for more immediate precursors to our mind wandering. Chimpanzees may well be able to remember where specific events occurred. One clever female chimpanzee has been taught to use visual symbols, known as *lexigrams*, to represent objects. These are displayed on a large screen, much as icons are shown on your iPad or iPhone. The chimp was able to select a lexigram for a given food item, and then point to where that item had been hidden some time beforehand. She could do this after delays of up to sixteen hours and indicate the hidden location to a person who didn't know the object had been hidden, let alone what the object was.[37] This is equivalent to naming a food item and

< 69 >

then showing its hidden location. Once again, though, this does not really distinguish knowing from remembering—the chimp may simply know where the item is without actually remembering the act of hiding it.

Perhaps the most eloquent advocate for animal thought and consciousness was the late Donald R. Griffin, whose 1976 book *The Question of Animal Awareness* claimed that animal communication offers "a possible window on the animal mind." The following anecdote from his 2001 book *Animal Minds: From Cognition to Consciousness* suggests episodic memory as well as a capacity for reasoning:

> A hungry chimpanzee walking through his native rain forest comes upon a large *Panda oleosa* nut lying on the ground under one of the widely scattered Panda trees. He knows that these nuts are much too hard to open with his hands or teeth and although he can use pieces of wood or relatively soft rocks to batter open the more abundant *Coula edulis* nuts, these tough Panda nuts can only be cracked by pounding them with a very hard piece of rock. Very few stones are available in the rain forest, but he walks 80 meters straight to another tree where several days ago he had cracked open a Panda nut with a large chunk of granite. He carried this rock back to the nut he has just found, places it in a crotch between two buttress roots, and cracks it open with a few well-aimed blows.[38]

Anne Russon and Kristin Andrews tell of a three-year-old orangutan who was observed trying to extract a small stone that had pierced the sole of her foot. One of the research assistants helped pick out the stone and added some latex from the stem of a fig leaf to help heal the wound. Eight days later, the animal approached the assistant, attracted her attention, seemingly acted out the leaf treatment, and showed that the wound had now healed. The assistant interpreted this as evidence that the orangutan remembered the treatment, and even several months later the animal would display her foot whenever she saw the assistant. This was also observed by one of

< 70 >

the authors of the article describing this behavior.[39] These examples are anecdotal, to be sure, but they do suggest episodic memory for particular events.

A more recent Japanese study with six chimpanzees and six bonobos seems to show quite detailed memory for single events. They were shown a movie in which a costumed King Kong came through one of two doors and attacked a human actor. When shown the same movie twenty-four hours later, they looked at the same door in anticipation of King Kong appearing. In another script, King Kong emerged but was then attacked by the actor with one of two objects which were located in different positions. On the second viewing, the locations of the objects were switched, and the animals looked at the particular object before the actor picked it up, even though it was now in a different position.[40] These results depend on the strong cuing of memory from the repeat of the movies and are in that respect not quite like the spontaneous recall of events from our own past lives, but they do suggest that great apes can recover information from a single episode in enough detail to anticipate what comes next and indeed to adjust to changes in the script. They suggest that these apes, our closest living nonhuman relatives, remember movies in much the same way that we do. Better than I do, probably, since I seem to forget most movies within hours—although King Kong does seem to linger.

What about imagining the future? Those cheeky scrub jays seem able to select between alternative choices of food in terms of anticipated hunger on a future occasion, even though satiated at the time of choice—much as satiated humans roam the supermarket in search of food for later meals.[41] Similarly, chimpanzees and their close relatives, bonobos, can save tools not needed in the present for use up to fourteen hours later, suggesting that they have in mind a future event.[42] The Swedish scientists Mathias Osvath and Elin Karvonen tell of a male chimpanzee in a zoo who collects caches of stones, which he later throws at visitors. He also cleverly hides these caches so that they will later be invisible to the unsuspecting visitors.[43] The stockpiling of weapons for future attack is not unknown to our own

< 71 >

species and seems to suggest planning of a future episode—even if the episode will involve defense rather than attack, as some of our leaders want us to believe.

The evidence for mental time travel in nonhuman species, whether in the form of remembered events or the planning of future ones, remains somewhat sporadic, and it is often possible to conjure up alternative explanations that don't require mental time travel itself. Animals may simply know about the outcomes of past events without mentally reliving those events, and they may learn to behave in ways that bring benefit in the future without actually imagining future events. Part of the difficulty here is that mental time travel implies subjective awareness, and we don't really have good access to what is going on in an animal's mind. Our normal access to the minds of others is through language, and animals don't appear to have the capacity to tell us what they're thinking about.

Nevertheless there is one avenue into the animal mind that does suggest that other animals may have wandering minds.

The Hippo in the Brain

First, some history. One of Charles Darwin's supporters was the English cleric, professor, historian, and novelist Charles Kingsley. In his 1886 book *The Water Babies* he gave the following advice: "You may think that there are other more important differences between you and an ape, such as being able to speak, and make machines, and know right from wrong, and say your prayers, and other little matters of that kind; but that is a child's fancy, my dear. Nothing is to be depended on but the great hippopotamus test."[44]

He was actually referring not to that large animal but to a structure in the brain known as the hippocampus minor. The nineteenth-century anatomist Richard Owen had maintained that this structure was unique to humans, rather as Descartes had earlier found human uniqueness in the pineal body. But Darwin's friend and advocate Thomas Henry Huxley disproved this by showing that all apes pos-

< 72 >

sess this structure, so that Owen's claim was without foundation. The dispute was the subject of a good deal of satire at the time, with *Punch* magazine weighing in with the following lines:

> Then HUXLEY and OWEN,
>   With rivalry glowing,
> With pen and ink rush to the scratch;
>   'Tis Brain versus Brain,
>   Till one of them's slain;
> By Jove! it will be a good match!
>   Says OWEN, you can see
>   The brain of a Chimpanzee
> Is always exceedingly small,
>   With the hindermost "horn"
>   Of extremity shorn,
> And no "Hippocampus" at all.
>   Next HUXLEY replies,
>   That OWEN he lies,
> And garbles his Latin quotation;
>   That his facts are not new,
>   His mistakes are not few,
> Detrimental to his reputation.[45]

The hippocampus minor soon disappeared from contention and is now known by its original name, *calcar avis*, meaning "cock's spur." It is now discussed only in the most obscure corners of anatomy texts and probably doesn't do much for us at all.[46]

More recent work has placed much more emphasis on the hippocampus major, now known simply as the hippocampus. It is a seahorse-shaped structure located in the inner surface of the temporal lobe—roughly behind your ears, if you need directions. Removal of the hippocampus, or severe damage to it, may cause the victim to be stuck in the present. One dramatic case is the English musician Clive Wearing, who suffered severe amnesia following a viral infection that destroyed part of his hippocampus in 1985. Deprived of the

< 73 >

memory of past episodes and the ability to imagine events that might happen in the future, he lives in the present, constantly under the impression that he has just woken up, or risen from the dead. His plight is vividly described in his wife Deborah's book *Forever Today* and in the United Kingdom's ITV television documentary *The Man with the 7-Second Memory*.

The most extensively studied case was Henry Molaison, a man long identified in the literature as H.M. He may well be the most famous patient in the history of neurology, and when he died in 2008 at the age of eighty-two, obituaries were published in the *New York Times* and the respected medical journal the *Lancet*. At the age of twenty-seven he underwent surgery for intractable epilepsy, and it was the surgery that was most responsible for destroying his hippocampus. He formed no memories back to the time of his operation and remembered little of his earlier life as well. His semantic memory, though, remained relatively unaffected; he remained able to talk normally, and his IQ was above normal. Like Clive Wearing, Henry seemed as incapable of imagining the future as of recalling the past.

Over the years after Henry's fateful operation, Suzanne Corkin carried out most of the testing on Henry and came to know him well. He failed to remember any of the many sessions he had with her, though, and always greeted her by recounting the same few stories that he could remember from his childhood. With curious insight, he once described his relationship with Corkin: "It's a funny thing—you just live and learn. I'm living and you're learning." Echoing the title of Deborah Wearing's book, Corkin published an account of Henry's life and psychological condition in 2013 under the title *Permanent Present Tense*.

The hippocampus, then, is the hub of the brain circuit involved in episodic memory and mental time travel. Brain imaging shows it to be activated both when people actively remember past events and when they imagine possible future ones.[47] It is also activated when people are asked to imagine purely fictitious episodes[48]—ones that might be prefaced with the phrase "Once upon a time." Other brain regions are involved, reflecting the fact that memory and imagination in-

< 74 >

volve information stored in widely dispersed areas, but the hippocampus appears to be the most critical component in that damage to it has the most debilitating effect on the ability to mentally escape the present. If mental time travel is unique to humans, then, the hippocampus may well contain something of the secret of what makes us human. Clive Wearing retains some human characteristics—he can engage with people, recognizes his wife and children, and can converse despite rapidly losing the thread of a conversation. He also retains something of his musicianship and can conduct a choir. But in losing access to the past or future, he has lost much of what it is to be a person.

But the default-mode network itself, responsible for our mind wandering, is identifiable in primates[49] and even in the rat[50]—and in rats, as in humans, the hippocampus plays a critical role in memory. Recordings from the hippocampus of the rat reveal that single neurons code where the animal is located in a spatial environment. These neurons are known as "place cells" and together generate what has been termed a "cognitive map" of the environment that tells the rat where it is.[51] Humans too—the hippocampus is enlarged in licensed taxi drivers in London, famous for knowing their way round that confusing city.[52] Bus drivers don't show the same enlargement but are better at learning new landmarks.[53] It seems it's best not to overload your spatial memory, especially if you plan to move to new territory.

That said, the hippocampus supplies us with a spatial sense that is remarkably flexible. I can locate myself in my study, where I now sit. Or I can imagine myself in another room—or in my office at work. The hippocampus even in the humble rat, it seems, can construct eleven different maps for each of eleven different rooms.[54] And as I explain below, the rat may have ability to rotate its mental map or zoom from local to more global representations. The hippocampus is one busy little seahorse.[55]

< 75 >

The Secret Life of Walter Ratty

Some of the activity in the hippocampus of the rat also seems to reflect mind wandering. Not only does activity in place cells in the rat hippocampus show where the rat is currently located in a maze, but it sometimes also shows activity with the same signature after the animal has left the maze. This activity consists of sharp-wave "ripples" superimposed on slow-wave activity when the animal is immobile or asleep, and has been interpreted as a speeded-up "replay" of earlier maze experience. What is interesting, though, is that the activity need not correspond to what the animal actually did. For instance, the rat may have actually traveled from A to B to C. Later activation, while the rat is outside of the maze, may correspond to the same path or the reverse path (C to B to A), or even a path elsewhere in the maze that the animal had not actually traversed (say D to E to F).[56] This activity seems to be involved in consolidating the memory for the animal's wandering in the maze, and also for constructing a cognitive map that includes the parts of the maze that the rat did not actually visit.[57]

In another experiment rats were trained to alternate left and right turns at a particular location in a maze. Between trials they were introduced to a running wheel, and while they were running, activity in their hippocampuses was recorded. This activity coded which way the rats planned to turn in the maze on the next trial. (My mind wanders, too, when I'm on a treadmill.) That is, the rats seemed to be planning ahead. The authors of this study write that autonomous activity in the hippocampus, "having evolved for the computation of distances, can also support the episodic recall of events and the planning of action sequences and goals."[58]

In a recent review, May-Britt Moser, David C. Rowland, and Edvard I. Moser write that "replay [in the rat hippocampus] can either lead or follow the behavior once the map of space is established. In that sense, the replay phenomenon may support 'mental time travel' . . . through the spatial map, both forward and backward in time."[59] They go on describe some work on human patients suggesting that sharp-wave ripples may indeed represent conscious

< 76 >

mental activity, whether in rat or in human. In one study, human patients about to undergo surgery had electrodes implanted in place cells in the medial temporal lobe, in an attempt to locate the source of epileptic seizures. They were given the task of navigating a virtual town on a computer screen and delivering items to one of the stores in town. They were then asked to recall only the items and not the location to which they were delivered.[60] The act of recall, though, activated the place cells corresponding to that location, effectively mirroring the replay of place cells in the rat brain.

And there may be more to Walter Ratty's mind wandering than just imagined trajectories in a maze. Moser and her colleagues summarize evidence that cells in the rat hippocampus respond not only to specific locations but also to features of environments they have explored, such as odors, touch sensations, and the timing of events. The human hippocampus no doubt has a richer capacity to add non-spatial information to our episodic memories than does the rat, whose social and practical world is relatively impoverished (or so we like to think), but the basis for our ability to replay episodes, and imagine future ones, may go well back over sixty million years to our common mammalian ancestry. Perhaps it goes back even further to our common ancestry with birds. There was once a view that birds do not possess hippocampuses, leading to an old joke that the purpose of the hippocampus was to prevent flight. This is wrong—they are as fully hippocamped as the hippopotamus, and of course birds need to be especially well informed about space and perhaps to the events that occurred in different places in the past. And anyway our national bird in New Zealand, the kiwi, doesn't fly.

The mapping of locations in space, whether in the present, future, or past, requires constant updating as the animal moves around. This is called remapping, and it must be fast and flexible. It appears to depend on an area of the brain called the entorhinal cortex, which lies between the hippocampus and the neocortex and acts as a gateway for the calibration of hippocampal mapping. It contains cells laid out in a grid pattern, and different grid-cell modules are dedicated to different geometric aspects of the environment and its relation to the

< 77 >

animal. These include its contours, the orientation of the animal's head in the environment, and the spatial scale. Within the hippocampus itself, different spatial scales are represented along the axis of the hippocampus, ranging from a more detailed, close view at the rearward end toward a broader, more distant view at the front end. In the rat, the rearward portion operates on a region of about 1 meter in width, while at the frontward the ventral end the width is about 10 meters.[61]

This kind of system allows you to zoom out or in to represent the different scales. I can ask you to locate yourself in the room where you are now, perhaps reading these very words, or I can ask you to zoom out so you locate yourself in your house, your suburb, your city, your country. I can also ask you to turn around so that the orientation of your spatial map is correspondingly altered. All of this can happen in the imagined past or future, as well as in the present; You might imagine, for example, a hotel room where you stayed in Paris, or zoom out to imagine where that hotel is located, or further to where Paris itself is located. These remappings can be almost instantaneous. The mechanisms for them appear to be present even in the rat brain, albeit no doubt in more restricted contexts.

Variations in scale may apply to our understanding of events in the world as well as simply to spatial awareness. In one study, people were shown sequences of four videos of different events, along with narratives describing the events. At one level narratives were linked to each video, encouraging attention to individual details. At the next level narratives linked a pair of videos, and at the final level a narrative linked all four videos. As the people processed these narratives, activation in the hippocampus progressed from the rearward end to the forward end as the scale of the narrative shifted from small and detailed to larger and more global.[62] This probably happens when you read a novel. Page by page, you focus on the details, but as the story progresses you build a progressively more global understanding of what the story is about. Be thankful to your hippocampus that you can make sense of a novel at all.

The combinations of just a few of the entorhinal grid modules

< 78 >

underlying aspects such as orientation and scale can generate a vast number of patterns of activity in the hippocampus, since each grid module can assume different levels. Moser and her colleagues write, "The mechanism would be similar to that of a combination lock in which 10,000 combinations may be generated with only four modules of 10 possible values, or that of an alphabet in which all words of a language can be generated by combining only 30 letters or less."[63] One cannot but be reminded of the Chomskian notion of discrete infinity.

Implications for Language

Although the generativity of spatial mapping is nonlinguistic, it may well underlie the generativity of language itself. Even the rat appears to be capable of constructing a potentially vast number of arrays of spatial maps, drawn from experience but projected into future possibilities and populated also with nonspatial elements. In the rat these elements may be restricted to simple aspects like sounds or smells, and we may perhaps allow ourselves the luxury of believing our own experiences to be incomparably richer. Yet the generative component itself probably has a long evolutionary history. As Darwin famously put it: "The difference in mind between man and the higher animals, great as it is, certainly is one of degree and not of kind."[64]

On this view, the generativity of what Chomsky calls I-language is a property not of language itself but rather of the thought processes that language is designed to communicate. Generative thought is universal to the extent that we all inhabit space and represent the objects and actions within it. The way in which events are represented, moreover, can be recursive in that we can zoom, locating scenarios within scenarios. These representations are not restricted to the present; we hold representations of past events and imagine future ones—and even fictional ones. Of course some of our thoughts are not spatial—we have emotions, opinions, and abstract ideas, but our ways of structuring them may well be built on our understanding of space. But where this account differs from Chomsky's is that he argues that

< 79 >

the generative, combinatorial nature of I-language evolved entirely within our own species, and perhaps even originated in a single individual and disseminated from there. The account given here suggests that the origins of I-language lay in mind-wandering capacities that go far back in the evolution of animals that move in space and need to know where they are, what has happened to them and where it happened, and what they might do next.

There remains, of course, the Rubicon—that apparent gap that separates speaking (or signing) humans from inarticulate apes. That gap, though, may depend on the ability to communicate our wanderings rather than the wandering themselves. The gap is one of externalization rather than internalization—E-language rather than I-language. Although apes and rats may be capable of a degree of mind wandering, they show little evidence that they can communicate about the nonpresent except perhaps in the limited context of pointing, as we saw in chapter 2. And as we also saw there, one of the fundamental properties of language is *displacement*—indeed, the linguist Derek Bickerton regards displacement, rather than arbitrariness, as the "road into language."[65] Displacement itself may be within the cognitive capacity of other species, maybe even rats, but we humans have evolved complex routines that enable us to share our mind wanderings in considerable detail, thereby gaining from the experiences of others. The problems inherent in communicating about our mind wanderings may have driven the shape of language itself, including the intricacies of grammar.

The question of how we communicate our mind wanderings is the topic of part 3. Before we delve into that, we need to consider other critical aspects of cognition that also have a bearing on how language is shaped.

< 80 >

5

MIND READING

Ferdinand and Oscar are given a cake to share. Ferdinand agrees to cut the cake and then proceeds to take the larger portion. When Oscar protests, Ferdinand says, "Well, if you had cut the cake, would you have taken the smaller portion?" Oscar agrees that he would have done so. "So," says Ferdinand, "you got the piece you wanted."

Ferdinand knows what's in Oscar's mind, although his own mind is more concerned with what he will put in his stomach. Our ability to understand what is in the minds of others is known as *theory of mind*. This depends little on language, which really only scratches the surface of underlying thought, as we saw in the previous chapter. Rather than theory of mind depending on language, language really depends on theory of mind. If we didn't know what was going on in the minds of others, we could not really have a meaningful conversation with them. Through language we might influence the flow of thought in another person, rather in the manner that one might steer a car through the streets of a city, but just as the car needs an engine, so the mind needs its own internal impetus.

The role of theory of mind in language was recognized by the philosopher Paul Grice, who held that the speaker not only must know what is in the listener's mind but knows that the listener knows this. Here's Grice's rather contorted way of putting it: "He said that P; he could not have done this unless he thought that Q; he knows (and knows that I know that he knows) that I will realize that it is nec-

< 81 >

essary to suppose that Q; he has done nothing to stop me thinking that Q; so he intends me to think, or is at least willing for me to think that Q."[1] If you're not used to minding your Ps and Qs, the basic idea is that Q is the underlying thought and P is what is said, and the listener knows that the speaker intends her to have that thought. Oh, and the speaker knows that the listener knows this. Got it?

Notwithstanding the complexity of his P-Q statement, Grice was also famous for his maxims for effective conversation. One of them, the Maxim of Quantity, urges that we make our contribution as informative as required but not *more* informative than required. For instance, you might meet a colleague at work one morning and say "Hey, what a game!" You know that she watched the rugby last night and would have appreciated its quality. If you are speaking to a less familiar colleague, though, you might offer more information: "Did you watch the rugby last night? Great game, wasn't it?" Or to a bemused visitor from the United States: "Rugby, you know, is our national passion. There was a game shown on TV last night, and if you happened to watch it you would have seen rugby at its best."

Just how specific we need to be, then, depends on the audience, and when Grice penned the P-Q sentence quoted above he no doubt had in mind the way philosophers think. Even as I write this paragraph I have in mind the minds of the audience I hope to reach. That audience, of course, includes you. I assume for a start that you understand English and that your understanding of the world is broadly the same as mine, so that the various illustrations I use will make sense. I nevertheless try to spell out the arguments more than I might do if communicating with a close colleague in the same field—such colleagues won't read the book anyway, because they know me and the stuff I write all too well. The unspoken thoughts that underlie language can be complex and deep. As the cognitive linguist Gilles Fauconnier put it, "When we engage in any language activity, we draw unconsciously on vast cognitive and cultural resources, call up models and frames, set up multiple connections, coordinate large arrays of information, and engage in creative mappings, transfers, and elaborations."[2]

Dan Sperber and Deirdre Wilson give an example of how com-

< 82 >

plex mutual understanding can build up prior to the utterance of any words:

> Suppose that Peter and Mary are walking in the park. They are en-
> gaged in conversation; there are trees, flowers, birds, and people
> all around them. Still, when Peter sees their acquaintance John in
> a group of people coming towards them, he correctly predicts that
> Mary will notice John, remember that he moved to Australia three
> months earlier, infer that there must be some reason he is back in
> London, and conclude that it would be appropriate to ask him about
> this. Peter predicts Mary's train of thought so easily, and in such a
> familiar way, that it is not always appreciated how remarkable this
> is from a cognitive point of view.[3]

So Peter might say to Mary something like "Hmm. Wonder what brought him back so soon." They both know exactly what he means and know that the other knows it too.

At the other extreme, there are occasions when someone says some-thing unexpected yet we do understand what they mean. In chapter 1 I mentioned visiting a publishing house in England and was greeted by the unexpected statement "Ribena is trickling down the chande-liers." Although taken aback, I did understand what was said. The speaker probably knew that I would know what ribena is and that a stately building such as his might have chandeliers. He would have tacitly assumed that we shared a good deal of information—including of course the fact that we spoke the same language.

Sometimes words seem scarcely necessary to maintain a conver-sation at all. Some years ago I went in to a pub in the northwest of England. The conversation among the hardened old-timers went like this:

> "Aye."
> Considered pause . . .
> "Aye."
> An even more considered pause . . .

< 83 >

"Yup."

Pause . . .

"Aye."

Pause with some nodding of heads.

And so it went on for a while. I think they were talking about me.

Unlike most animal communication, human language is characterized by what has been called *underdeterminacy*. The words themselves are not enough to specify the precise meaning. Even very simple sentences can have very different meanings, depending on the context, and can be understood only if the communicator and recipient are in the same state of mind. One well-worked example is the sentence "Time flies." This might be taken to refer to the impression that events are taking place too rapidly, and perhaps uttered as an explanation for not getting a job done or with the realization that one is growing old too rapidly. Or perhaps it might be uttered in the context of an experiment on flies and is a directive to record the time it takes for flies to proceed from one destination to an attractive food source. Then again, an exasperated employee of *Time* magazine Australia might be compelled to expostulate on the annoying flies that buzz around her head.[4]

In his book *Speaking Our Minds*, Thom Scott-Phillips argues that human language was an entirely new form of communication in which "underdeterminacy is an inherent and inevitable feature."[5] It is true that this sometimes creates ambiguity and misunderstanding, but by the same token it also allows for the flexibility and creativity of language. And it can work only if we know what's going on the mind of the audience we are addressing and the audience knows reciprocally what's on our mind. This need not involve language itself. Scott-Phillips gives the example of being in a coffee shop, catching the attention of a waiter, and tilting his coffee cup in a stylized way. The waiter comes over and fills his cup. Here the gesture is underdetermined, but the thirsty customer and the waiter both know what's in the mind of the other. Scott-Phillips refers to this kind of communication as *ostensive-inferential*—ostensive because it involves an act of

< 84 >

showing and inferential because it require an act of interpretation.[6] We carry out many such acts, such as shrugging, pointing, raising eyebrows, but the most complex form is language itself, whether spoken or signed.

One reason that humans are blessed with language, and other species are not, then, could be that we alone are capable of theory of mind, at least to the required level. Nonhuman animals may have relatively little notion of what is going on in the minds of other animals and perhaps no reason in any case to share their thoughts. So let's consider what we do know about theory of mind in animals.

Theory of Mind in Animals

First, there seems little doubt that animals are capable of reading emotion in others, just as we humans are. In Shakespeare's *Henry V*, act 3, scene 1, the King urges his troops to

> imitate the action of the tiger;
> Stiffen the sinews, summon up the blood,
> Disguise fair nature with hard-favour'd rage;
> Then lend the eye a terrible aspect.

The enemy, it is assumed, will read the emotion, just as the antelope may read the emotion of the marauding tiger. In his book *The Expression of the Emotions in Man and Animals*, Charles Darwin gives detailed examples of the ways fear and anger are expressed in cats and dogs and does not neglect the expression of positive emotion, such as I hope the reader of this book might display to those around her: "Under a transport of Joy or of vivid Pleasure, there is a strong tendency to various purposeless movements, and to the utterance of various sounds. We see this in our young children, in their loud laughter, clapping of hands, and jumping for joy; in the bounding and barking of a dog when going out to walk with his master; and in the frisking of a horse when turned out into an open field."[7]

< 85 >

The primatologist Frans de Waal points out that animals not only detect emotion in others but also show empathy, a trait that goes far back in evolution to the need for maternal care.[8] Mice as much as elephants must be in tune with signs of hunger, pain, discomfort, or danger in their young—their survival, and indeed that of the species, depends on it. Animals also respond with empathy to perceived distress in others, as several experiments have shown. Mice perceiving pain in other mice intensified their own reaction to pain;[9] monkeys refused to pull a chain to receive food if doing so caused a shock to be delivered to another monkey.[10] Chimpanzees sometimes behave to ensure the satisfaction of others. In one study, female chimps were tested in pairs, with one chimp being offered tokens that could be exchanged for food reward. Tokens of one color, say green, brought reward to the chimp that made the choice, while tokens of another color, say green, brought equal reward to both the chimp that chose and her partner. They showed a bias for the second option, even though there was no personal advantage in doing so.[11] This suggests a concern for the passive partner's satisfaction—or perhaps the chooser was scared of being beaten up later if she didn't ensure equal shares.

De Waal observes that apes, but not monkeys, will try to soothe a distressed animal, often with a consoling embrace. Bonobos, he suggests, are more empathic than the more aggressive chimpanzees. We are equally related to both, and it is a moot point which is the more powerful legacy. For a more peaceable world, we should strive to find our inner bonobo. Either way, we humans do show empathy and cooperation, in spite of occasional barbaric acts of cruelty, and we do so from a very early age. Even in the first year of life, infants will not only comfort those who are hurt but also help others do things by fetching objects that are out of reach or removing obstacles. Two-year-olds are gratified not only by helping others but also by watching others being helped by a third party. You can see it in their eyes.[12]

In apes, though, theory of mind may be restricted largely to the reading of emotion, and this perhaps is most evident in the nature of their communications. In humans, language reveals the thoughts of others, but animal cries and calls, even those of apes, seem restricted

< 86 >

to emotional communication. These calls are largely automatic, controlled by the limbic system—sometimes called the emotional brain. The fact that they are automatic rather than voluntary is probably adaptive, because it means they can't be faked. The roar of the lion and the mew of a kitten are honest signals. In humans, too, laughing and crying are not easily faked, although skilled actors can sometimes do a creditable job. Language itself is a different story, as well calibrated for lying and deception as for revealing truth, although body language can sometimes give the game away. As the American actress Mae West said, "I speak two languages, Body and English."

Of course manual gestures in apes have a more voluntary aspect and may indeed be the precursors to language, as I suggested in chapter 4. Actions like grooming or the "directed scratch" suggest voluntary intent and perhaps some awareness of what's in another animal's mind. Even so, such actions seem more emotional than is evident in the dispassionate way we humans talk to each other. Of course people do rant and rave, and may even demand to be scratched, but language can convey complex information, like instructions on how to bake a cake or descriptions of what one did on a vacation, without requiring the stimulus of emotion.

Nevertheless, the absence of humanlike language in apes need not mean that they have no ability to read the minds of others. Perhaps weary of the seemingly futile search for languagelike abilities in apes, David Premack and Guy Woodruff, in a seminal paper published in 1978, asked the question "Does the chimpanzee have a theory of mind?"[13] One approach they tried was to show the chimpanzee Sarah videos of a human grappling with some problem and then offer her a choice of photographs, one of which showed a solution to the problem. In one example they showed a woman trying to escape from a locked cage, and one of the photos showed a key, along with other objects irrelevant to the task. Sarah performed quite well at appropriately choosing the key, although, as Premack and Woodruff recognized, this need not show that she appreciated what was going on in the mind of the person depicted. For example, the key might have been selected through simple association with the cage.

< 87 >

A different approach is to test whether a chimpanzee might understand what another individual can see—whether it has a theory of the "seeing mind." Daniel Povinelli noted that chimpanzees readily approach humans to beg for food, offering an opportunity to test whether they are influenced by whether the human can see or not. Given the choice of two individuals to beg from, one with a blindfold over her eyes, chimpanzees failed to systematically choose the one who could see. The same was true when one of the individuals had a bucket over her head or covered her eyes with her hands.[14] This seemed to suggest that chimps don't understand whether a person can or cannot see. (It seems something of an irony that humans should be asking whether a chimp can read a human mind, when we humans go to elaborate lengths to figure out how to read the chimp mind.)

Other findings, though, suggest that chimps do sometimes understand what is visible to a human. In one study the experimenters arranged for a human to compete with chimpanzees for access to items of food. Faced with this challenge, the chimps chose to approach the food via a route hidden from the human's view.[15] The route taken was sometimes quite circuitous. Chimps may well understand when something is in a person's view without fully appreciating that a person's view is blocked when the eyes are covered.

Chimps may not always discern what a person can see but may be rather more adept at understanding what another *chimpanzee* can see. A chimp approached food when a more dominant chimp could not see the food, but was reluctant to do so when it could see that the food was visible to the dominant chimp.[16] The same seems to be true of marmosets, which preferred to choose food that a dominant marmoset couldn't see.[17] Again, a subordinate chimp retrieved hidden food if a dominant chimp was not watching while the food was hidden, or if the food was moved to another location while the dominant chimp wasn't watching. The subordinate also retrieved food if a dominant chimp watched it being hidden but was then replaced by another dominant chimp who hadn't watched, suggesting the subordinates could keep track of who knew what. The subordinates did not

< 88 >

succeed on another test, though, in which a dominant chimp watched one piece of food being hidden but did not watch another piece being hidden; they failed to consistently choose the food that the dominant chimp hadn't watched.[18] Perhaps it all gets a little confusing, like trying to remember who likes whom when you plan a dinner party.

Chimpanzees, it must be said, do pose something of a dilemma for us insecure humans. On the one hand, they and their cousins the bonobos are our closest living nonhuman relatives, but on the other hand we have an all-too-human need to declare our superiority, especially in intellectual or moral terms. Perhaps it is not surprising, then, that even more than thirty years since Premack and Woodruff posed the question, there is still sharply divided opinion as to whether chimps can truly read the minds of others. In an article titled "Darwin's Mistake: Explaining the Discontinuity between Human and Nonhuman Minds," the trinity of Derek Penn, Keith Holyoak, and Daniel Povinelli conclude that even chimpanzees have no theory of mind[19]—and that Darwin was wrong in declaring the difference between humans and nonhumans to be one of degree but not of kind. Josep Call and Michael Tomasello conclude, more generously, that the thirty years of research have shown chimpanzees to have an understanding of the goals, intentions, perceptions, and knowledge of others but no understanding of others' beliefs or desires.[20] It seems that we humans, so proficient at reading each other's minds, are not so good at reading the mind of the chimpanzee. Nevertheless, a Darwinian perspective favors Call and Tomasello over the bleakness of Penn and his skeptical colleagues.

At least some forms of communication between chimpanzees might be taken to imply reciprocal theory of mind. In a field study, Katie Slocombe and Klaus Zuberbühler report that the screams of wild chimpanzees under attack were modified depending on the audience.[21] The scream would be intensified if another chimpanzee was of equal or higher status than the attacker, presumably in the hope that this chimp might come to the rescue. This might be taken to imply that the victim understood that her predicament would be understood by the proposed savior. Perhaps better examples might be

< 89 >

found in gestural communication among apes, such as the "directed scratch" or the many playful gestures recorded by Catherine Hobaiter and Richard Byrne in a study of chimpanzees in the wild in Uganda, as described in more detail in chapter 7. As these authors point out, their gestures do seem intentional and referential and are often loose and seemingly ambiguous. In these respects they may well meet the definition of what Thomas Scott-Phillips calls *ostensive-referential*— the underdetermined system of communication that is a precursor to language.

Even so, both human theory of mind and human language seem vastly more sophisticated than their equivalents in chimpanzees. But the difference may still be one of degree. Theory of mind is potentially recursive, in that I may know not only what's in your mind but also that you know what's in my mind. In the Gricean view, as also developed by Scott-Phillips, it is this extra level of recursion that is necessary for true language. If two people are to have a meaningful conversation, each must not only know what's in the other's mind but also know that the other knows this.

To put it another way, language is a mechanism for the sharing of knowledge or experience, and sharing implies reciprocity. The sharing of information in the developing child is first evident in pointing. Young children point at objects, not simply because they want them but because they want to share their appreciation with you. This continues in their early language. A few years ago I occasionally had the privilege of picking up my granddaughter from daycare. From the age of two she constantly pointed at things she saw—horses, a train, a helicopter. She did not want these things; she simply wanted to share the fact that she had seen them. Chimpanzees, in contrast, do not seem to share in this way. When they point, it is to request something, such as food that is out of reach. They point mainly to achieve their own gratification. This may reflect a first-order theory of mind, in that they may know that their minder can see the object in question, but there is no evidence that they know that the minder knows that they know this.

The question, then, is why did humans or our hominin predeces-

< 90 >

sors develop a deeper level of social awareness—a level that may in fact have been a necessary precursor of language? The answer probably lies not in some sudden genetic mutation but rather in the gradual emergence of increasingly complex social behaviors. They acquired what has been called *social intelligence*.

Social Intelligence

As primates, our forebears had long existed as forest-dwelling animals, with limbs and bodily stature adapted to living in trees. With deforestation, they were forced to adapt to a more open terrain. A modern study suggests that there were two periods of deforestation, both occurring after our hominin forebears separated from the line leading to modern chimpanzees and bonobos.[22] The first occurred from about 4.5 to some 6 million years ago, corresponding to the appearance of the earliest known bipedal hominins, *Ardipithecus ramidus* and *Ardipithecus kadabba*. But then the African environment became more wooded again during the time of the successors to *Ardipithecus*, the australopithecines, for about the next 1.5 to 2 million years. Deforestation increased again from around 2.9 million years ago and persisted through the Pleistocene, now dated from 2.88 million years ago to 12,000 years ago.

This second period of deforestation saw the emergence of the genus *Homo*, leading finally to our own triumphant species, *Homo sapiens*. This period was especially critical to the emergence of social intelligence, not least because of the presence of large carnivorous animals, whose numbers peaked in the early Pleistocene. They included at least twelve species of saber-tooth cats and nine species of hyena.[23] Our puny forebears had previously been able to seek cover from these dangerous predators in more forested areas, and perhaps earlier by retreating into water as suggested below, but such means of escape were relatively sparse on the savanna. Danger lurked even in the water, just as today people are killed in shark attacks or have unpleasant experiences with stingrays or unfriendly jellyfish.

< 91 >

The solution seems to have been the emergence of the "cognitive niche," alluded to in chapter 3. Our forebears sought safety in numbers and social cooperation. The size of the group grew, leading to increased social complexity. As the evolutionary anthropologist Robin Dunbar has emphasized, the number of relationships between individuals in a group increases much more rapidly than the size of the group itself, so the groupings evolved complex structures. Some of this structure was hierarchical, with individuals nested in families, extended families, villages, tribes, clans—and eventually, and not always happily, nations. Based on the relation between group size and brain size, Robin Dunbar has suggested that humans are adapted to live in functioning groups of around 150—the so-called Dunbar number.[24] The structuring of groups was also horizontal, with different subgroups becoming specialized for different activities and skills—hunting, making weapons, preparing food, educating children . . . In the increasingly structured social world, theory of mind—the ability to understand how and what others think—would be critical to stability and effective cooperation.

Some of the advantages of cooperation are illustrated in Brian Skyrms's treatise *The Stag Hunt*. A hunter working alone might be able to kill a rabbit, but two hunters working together might be able to bring down a stag, which is worth more than twice a couple of rabbits. But this requires theory of mind. Two hunters who know each other well and read each other's minds would undoubtedly hunt more effectively than a pair of strangers working together. Enhanced communication skills would also have been necessary to allow individuals to report back to the group on where there might be good grounds for hunting or fishing.

Cooperation seems to be instinctive. David G. Rand and colleagues at Harvard University arranged people in groups of four and set them a task of deciding how much money to give to a common project. The amount they gave was greater the more quickly they made their decision.[25] This seems to suggest that your immediate intuition is to be cooperative but that the more you think about it, the less cooperative you become. This is perhaps why Caesar, in Shakespeare's

< 92 >

Julius Caesar, complains, "Yon Cassius has a lean and hungry look. He thinks too much. Such men are dangerous" (act I, scene 2).

Our more considered decisions, which often override our earlier impulses, may well have an evolutionary basis too. Through cooperation and enhanced social intelligence, our forebears seemed to have conquered the threats posed by the inhospitable environment but then discovered a further threat—themselves. Richard D. Alexander writes, "The real challenge in the human environment throughout history that affected the evolution of the intellect was not climate, weather, food shortages, or parasites—or even predators. Rather, it was the need to deal continually with our fellow humans in social circumstances that became ever more complex and unpredictable as the human line evolved."[26] Once you can read the minds of others, you have the opportunity to deceive. For instance, if I know that you think I'm honest, I can perhaps ask you for a loan, which I promise to pay back. I then skedaddle with the loot. But then I may know that you think I know that you think I'm honest, so you might become suspicious of giving me the loan. Deception and intrigue can be pushed to ever deeper levels of recursion, in a spiral of intrigue.

Nicholas Humphrey attributes to Niccolo Machiavelli the aphorism "It is double pleasure to deceive the deceiver" and cynically describes the Machiavellian loop as a "self-winding watch to increase the general intellectual standing of the species."[27] But Humphrey was himself deceived, because the quote comes not from Machiavelli's *The Prince* (as widely assumed) but from the poem-fable "The Cock and the Fox" by the seventeenth-century French poet and fabulist Jean de la Fontaine. The philosopher Kim Sterelny suggests, though, that the Machiavellian loop may be more a product of modern mass society than of the smaller and more intimate social groupings of our forebears.[28] Mass society is anonymous, and con men can disappear after they've done their nefarious deed, and the problem may be exacerbated in the age of the Internet. Beware of e-mail promises of unexpected legacies from Nigeria. Still, the recursive intricacies of human mind reading apply to a social understanding that need not imply malicious intent. David Premack mischievously comments on

< 93 >

human sexual politics: "Women think that men think that they think that men think that women's orgasm is different."[29] (And it's a man, Premack, who thinks *that*.) And of course our social lives are fraught with dizzying recursion. Here's how the literary scholar Brian Boyd summarizes the plot of *Twelfth Night* by that fine old social commentator William Shakespeare:

> Maria foresees that Sir Toby will eagerly anticipate that Olivia will judge Malvolio absurdly impertinent to suppose that she wishes him to regard himself as her preferred suitor.[30]

Our social lives are a tangled web, whether or not we practice to deceive.

Perhaps the best indicator of the flowering of the social mind is brain size, which increased nearly threefold through the Pleistocene. This in turn no doubt depended on a succession of genetic changes. One gene of special interest delights in the aristocratic name Slit-Robo RhoGTPase-activating protein 2, or SRGAP2, and its influence on brain size seem to have depended on its being partially duplicated in the human genome, so that humans uniquely possess four different copies. From the parent gene, one duplication occurred about 3.4 million years ago, another about 2.4 million years ago, and another about 1 million years ago. The second of these corresponds roughly to the transition from the earlier australopithecines to the genus *Homo* near the beginning of the Pleistocene, signaling the beginning of the increase in brain size and the rapid advance of social intelligence. These duplications appear to have been carried by Neandertals and the recently discovered Denisovans, as well as by modern humans.[31]

On the Beach

One hypothesis is that our swollen brains may have come about through a period during which our forebears adapted to a semiaquatic existence foraging in water, along coastlines or by rivers or

< 94 >

lakes. This idea, known as the *aquatic ape hypothesis* (AAH), was first proposed by the German pathologist Max Westenhöfer in 1942 and then by the distinguished British biologist Sir Alister Hardy in 1960. Since then it has been enthusiastically pursued by Elaine Morgan, who has written several books on the topic, including her best seller *The Aquatic Ape*. Superficially, at least, it seems to explain a number of the characteristics that distinguish humans from other living apes. These include hairlessness, subcutaneous fat, bipedalism, our large brains, and even language.

But the idea of an earlier aquatic existence has probably gained more acceptance in the popular press than in the scientific literature, perhaps in part because of Morgan's feminist orientation and in part because the aquatic hypothesis runs counter to more widely accepted explanations of human evolution, such as adaptation to the savanna. Claims about the timing of the supposed aquatic phase have also been at odds with what is known about our evolutionary past. When Hardy wrote, it was generally assumed that our forebears split from the great apes over ten million years ago and that our forebears were australopithecines. It is now widely believed that the split occurred some six million years ago, and we have a more detailed understanding of the different stages of hominin evolution since that time.

The aquatic hypothesis has been refined over the years and has gained greater plausibility. There is still of course resistance, but among those who have endorsed it is the distinguished South African archaeologist Philip Tobias. Tobias died in 2012, but in a chapter published in 2011 he wrote: "In contrast with the heavy, earth-bound view of hominin evolution, an appeal is made here for students of hominin evolution to body up, lighten and leaven their strategy by adopting a far great emphasis on the role of water and waterways in hominin phylogeny, diversification, and dispersal from one water-girt milieu to others."[32]

So let's body up and plunge on.

In an extensive discussion of the aquatic ape hypothesis and the various controversies surrounding it, Mark Verhaegen suggests that our apelike ancestors led what he calls an aquarborial life, on the bor-

< 95 >

ders between forest and swamp lands.[33] There they fed on shellfish and other waterborne foods, as well as on plants and animals in the neighboring forested area. The critical adaptations that gave rise to our more distinctive aquatic characteristics, such as greatly diminished hairiness, developed later, during the Pleistocene, from nearly three million years ago. This signaled the arrival of *Homo*, the genus that led eventually to our own species, *Homo sapiens*. Others of the genus, including *Homo habilis* and *Homo erectus*, as well as our large-brained cousins the Neandertals and Denisovans, did not survive.[34]

In this view, the environment that first shaped our evolution as humans was not so much the savanna as the beach. During the Ice Ages the sea levels dropped, opening up territory rich in shellfish but largely devoid of trees. Our early Pleistocene forebears dispersed along the coasts, and fossils have been discovered not only in Africa but as far away as Indonesia, Georgia, and even England. Stone tools were first developed not so much for cutting carcasses of game killed on land as for opening and manipulating shells. Bipedalism too was an adaptation not so much for walking and running as for swimming and shallow diving. Verhaegen lists a number of features that seem to have emerged only in Pleistocene fossils, some of which are present in other diving species but not in pre-Pleistocene hominins. These include loss of fur; an external nose; a large head; and head, body, and legs all in a straight line. The upright stance may have helped individuals stand tall and spot shellfish in the shallow water. Later in the Pleistocene, different *Homo* populations ventured inland along rivers and perhaps then evolved characteristics more suited to hunting land-based animals. The ability to run, for instance, seems to have evolved later in the Pleistocene. But Verhaegen suggests that, in fact, we are poorly adapted to a dry, savannalike environment and retain many littoral adaptations (that is, adaptations to coastal regions): "We have a water- and sodium-wasting cooling system of abundant sweat glands, totally unfit for a dry environment. Our maximal urine concentration is much too low for a savanna-dwelling mammal. We need much more water than other primates and have to drink more

< 96 >

often than savanna inhabitants, yet we cannot drink large quantities at a time."[35]

Part of the reason for our swollen brains, may derive from a diet of shellfish and other fish accessible the shallow-water foraging. Seafood supplies docosahexaenoic acid (DHA), an omega 3 fatty acid, and some have suggested that it was this that drove the increase in brain size, reinforcing the emergence of language and social intelligence. Michael A. Crawford and colleagues have long proposed that we still need to supplement our diets with DHA and other seafoods to maintain fitness.[36] Echoing Crawford, Marcos Duarte issues a grim warning: "The sharp rise in brain disorders, which, in many developed countries, involves social costs exceeding those of heart disease and cancer combined, has been deemed the most worrying change in disease pattern in modern societies, calling for urgent consideration of seafood requirements to supply the omega 3 and DHA required for brain health."[37]

In other words, eat more fish.

Patterns of Development

Social intelligence is not just a matter of brain size itself; it is also linked to how the brain grows. The brain of the human fetus is larger than that of other primates, but bipedalism in humans narrowed the birth canal, creating what has been called an "obstetrical dilemma." The solution was to defer some of the growth of the brain until after birth—only a partial solution, perhaps, because birth is difficult enough anyway, as any mother can attest. Delayed growth of the brain lowered the metabolic resources available for the growth of the rest of the body, further increasing the helplessness of the human infant and the need for maternal care. Among other primates the infants cling to their mothers' hair, but human babies need constant attention and nurturance. Among hunter-gathering humans, mothers seldom put their babies down, and when they do it is only for a few

< 97 >

seconds. The duration of infancy is about the same in humans as it is in some other primates, but it is much more socially intense, and the social stimulation takes place in human babies while the brain is still developing and at its most impressionable. This sets the stage for such social skills as theory of mind and language.

There's more. Primates, including apes, skip from being infants to being juveniles, but in humans an extra stage, called childhood, intervenes. It runs from about age three to about age eight. During this period the rate of growth tapers off; the child becomes less dependent on the mother and begins to establish social relationships outside the family. According to John L. Locke and Barry Bogin, the insertion of childhood into the developmental sequence probably first appeared early in the Pleistocene in a new genus, *Homo*, of which *Homo sapiens* is the only surviving species.[38] Language continues to develop, to the point that children can begin to establish a knowledge-based language system. In most societies, children begin formal schooling at age five or six. As A. A. Milne put it in his poem "Now We Are Six":

> When I was five,
> I was just alive.
> But now I am six,
> I'm as clever as clever.
> So I think I'll be six
> Now and forever.[39]

Six is also the age at which the vocal tract is said to have matured to adult proportions, and spoken languages learned after that age are afflicted with accents, as earlier languages interfere. If you are moving to another country and want to avoid a foreign accent, it will be ever so clever to go before you are six.

Delayed growth of the brain extends even beyond childhood. Locke and Bogin document evidence that a further developmental stage, called *adolescence*, was inserted between the juvenile stage and adulthood, and is unique to humans and apparently not evident in earlier species of *Homo*—although they don't mention Neandertals. It

< 98 >

runs from about age eleven to about eighteen, or even later in under-graduate students. Although its main function seems to be to torment parents, it was no doubt important in fostering the transition to adult independence. In Shakespeare's *The Winter's Tale*, the Old Shepherd remarks "I would there were no age between sixteen and three-and-twenty, or that youth would sleep out the rest; for there is nothing in the between but getting wenches with child, wronging the ancientry, stealing, fighting" (act 3, scene 3).

And even then growth may not be complete. The white matter of the mature brain is made up of neurons that are coated in myelin, which speeds transmission from one part of the brain to another. In human newborns there is no myelin, whereas in chimpanzees some 20 percent of myelination is achieved before birth. Thereafter myelination proceeds much more slowly in chimpanzees and is complete by about age eleven, but in humans it continues into the thirties.[40] In the 1960s a familiar watchword among rebellious students was "Don't trust anyone over thirty." As age and myelination have colonized my brain, it seems to me that it's the under-thirties we should be wary of.

The emergence of social intelligence, at least in its complex human form, was therefore gradual, shaped largely during the Pleistocene. The slowed growth of the brain was adaptively calibrated to absorb the complexities of human social life. It seems fair to conclude that language, as an important component of our social intelligence, also evolved over this extended period and was not a result of a miraculous mutation within the past one hundred thousand years. As I suggested in chapter 2, language is part of the glue that binds us together—but more of that in the next chapter.

In this chapter and the last, I have tried to show the major ways in which experience itself evolved and expanded—through mental wandering in space and time and through appreciation of what goes on in the minds of others. These attributes are distinctively human, although probably not uniquely so. Behavioral and neurophysiological evidence suggests that mental precursors probably did exist in our forebears, perhaps even going back to our common ancestry with

< 99 >

mammals and even birds, but more developed in our closest great ape cousins. The integration of our wanderings in space and time, and even into the minds of others, forms the basis of that most human of activities, the telling of stories—of which even this book is an example. That is the topic of the next chapter.

< 100 >

6

STORIES

To say that all thinking is essentially of two kinds—reasoning on the one hand, and narrative, descriptive, contemplative thinking on the other—is to say only what every reader's experience will corroborate.

William James[1]

Narrative does indeed seem to form a large component of our thinking and may well have been the initial spur to language. It is evident not only in our fondness for telling stories but also in our obsession with TV soaps, movies, play acting, and everyday gossip. The other kind of thinking noted by James is reasoning, which may well have emerged later, perhaps as a result of the progression of language toward more abstraction, and I will touch on this later in the book. We normally think of narratives, or stories, as *told*, but the focus of this chapter is on the stories themselves and not on the telling of them. In this respect, then, I will treat stories as a mode of thinking, the "narrative, descriptive, contemplative" mode identified by William James. Of course our access to stories is typically through the telling—neuroscience has not yet developed to the point that we can read the minds of others in narrative detail! And even the telling of a story omits much of the underlying experience, as I suggested in the previous chapter.

Stories are built on the mental capacities documented in the previous two chapters, namely the capacity to travel mentally in time and space and into the minds of others, along with a dose of imagina-

< 101 >

tive construction. They take us on our journeys into often fictitious realms of existence, whether for instruction, for expansion of experience, or simply for entertainment. Some of our stories are simple replays of extended memories. We regale others, often tediously, with our recent travels, our latest golf round, or the exploits of our children—some of which are often better left untold. We also "narrativize" our plans for the future, the brilliant career that lies before us or the plan of attack in a coming match or business venture.

Katherine Nelson, writing of storied thoughts, concludes that "narrative models and early influences in early childhood help to transform the episodic memory system into a long-lasting autobiographical memory for one's own life, and thus a self-history, which to a large extent underlies our concept of self."[2] Narrative is not a slave to memory, though, and as we saw in chapter 4, activity in that hub of memory, the hippocampus, is evoked not only by recalling the past and imagining the future but also by imagining purely fictitious scenes. The dividing line between memory and fiction is indeed blurred; every fictional story contains elements of memory, and many memories contain elements of fiction. Stories may be set in the past, as in historical fiction, or in the future, as in prophetic books such as George Orwell's *1984* (a date now thankfully in the past), or they may be set in some unspecified period. Of course stories also carry their own ribbon of time. Relative to any given point in a novel, there is a past and a future, although the teller of a story has the luxury of being able to flit back and forth between them at her whim and to carry the reader with her. Within our stories, as within our memories, though, time moves forward. We may flit back to some past episodes in our lives, but these episodes are then played forward.

The extension and embellishment of episodic memories into stories may well be what does distinguishes us from other species, although who really knows what internal narratives might lie beneath the soulful eyes of a mute chimpanzee? Our predilection for stories is so ubiquitous that the literary scholar John Niles suggested our species should be renamed *Homo narrans*—the storytellers. Similarly, the Israeli historian Yuval Noah Harari, in his book *Sapiens:*

< 102 >

A Brief History of Humankind, goes so far as to suggest that it was fictitious stories that drove the evolution of language itself: "The truly unique feature of our language is not its ability to transmit information about men and lions. Rather, it's the ability to transmit information about things that don't exist at all. As far as we know, only Sapiens can talk about kinds of entities that they have never seen, touched or smelled."[3]

Aristotle[4] wrote that poetry is more important than history because it deals with possibility rather than with what has actually happened, and therefore stretches the mind—just as physical exercise stretches the muscles and adds to our well-being. By "poetry" he meant fiction, or works of the imagination, which can of course include some of what we now understand as poetry but also novels and plays, and had Aristotle been alive today he would undoubtedly have included movies and television soaps. He might have even watched them.

Fiction is in essence a simulation of the social world, in which the characters have minds with which the reader or watcher can identify. Immersed in fiction, you mentally enter the social world created by the novelist or playwright, just as in real life you enter the social lives of the people around you. The psychologist Keith Oatley writes as follows:

> As we read a novel or watch a film, we set aside our own goals and plans, take the goals and plans of a fictional character, and insert them into our planning processor, the device by which we usually construct our own planned actions in the world. The fictional work gives us cues as to what happens when each action is performed, and we then experience empathetically (within the simulation of the social world that we are running) the emotions that we would feel in relation to the outcomes of actions as depicted by the author.[5]

It is sometimes suggested that fiction is mere fantasy, a way of escaping from the rigors of real life, and is therefore to be discouraged. The evidence suggests, though, that it enhances empathy and

< 103 >

theory of mind, making us better able to understand others. Oatley and colleagues measured the amount of fiction and nonfiction that people read, and found that tests of empathy correlated positively with the amount of fiction read, but negatively with the amount of nonfiction.[6] Better a bookworm than a nerd if you want to get on in the social world. A meta-analysis has also shown an overlap between areas of the brain activated by reading narrative stories and those involved in theory of mind.[7] When I was a graduate student, Donald Hebb, the esteemed Canadian psychologist and neuroscientist, used to tell us that we could learn more practical psychology from reading novels than from reading the journals of experimental psychology. We'd have more fun too, I might add.

The literary scholar Brian Boyd argues that stories originate in play, and indeed we refer to theater productions, at least, as "plays." Play has a long evolutionary history. In a physical sense, mammals and birds indulge in play, typically in the form of rough-and-tumble, or play-fighting. Play-fighting can perhaps be regarded as a preparation for actual fighting, a testing of strength and perhaps also an exercise in bonding. Nevertheless play-fighting is clearly different, and animals typically give signals to indicate that they intend to play rather than fight seriously, or play rather than mate. Dogs, for example, have a characteristic "play bow," in which they crouch on the forelimbs while leaving their hind limbs straight and wag their tails. My granddaughter is more direct; she simply says "Play with me." Increasingly her play involves stories, whether told or acted out.

Young children seem to spend most of their time in play, reinforced by stories or television cartoons. Before going to school, they go to play centers. They seem to especially relish stories with elements of danger, as though playful exposure to fearful events might increase the ability to cope with true danger later in life—perhaps preludes to an obsession with crime fiction. The world of the three-year-old is one in which fact and fantasy are blended, and there is always the possibility that the Big Bad Wolf, or some other fearsome creature, is lurking behind a tree—or worse, under the bed. Children's nursery rhymes and fairy stories, from "Jack and Jill" to "Little Red Riding

< 104 >

Hood," tell of frightening events and disasters, usually (but not always) narrowly averted. Stories, then, provide a means of stretching and sharing experience so that we are better adapted to possible futures.

The natural tendency of children to engage in playful fantasy also seems directed at establishing appropriate modes of action in society. Young children act out some of the scripts of life, such as feeding a baby, doing a medical checkup, cooking a meal. "You be the pupil and I'll be the teacher," my granddaughter commands me in preparation for a scenario she wants to play out. The Russian psychologist Lev Vygotsky gives examples of two sisters, aged five and seven, actually playing at being sisters, each acting out what she thinks a sister should be rather than simply being a sister. Another common example is of a child imagining himself (*sic*) to be the mother and the doll to be the child, or even the child imagining being the child and having the mother play the role of mother! As Vygotsky remarks, these examples were "playing at reality."[8] An important aspect of these scenarios is the learning and application of the rules that operate in society. Vygotsky also suggests that play is a way of delaying gratification, so that the rules of play take precedence over more immediate desires: "The essential attribute of play is a rule that has become a desire."[9]

Of course play, whether physical or mental, persists into adulthood. Boxing, wrestling, and football in its various forms are examples. Again, the rules are all-important, serving as constraints against impulse and as ends in themselves. Indeed many aspects of sports are unpleasant and painful, but these aspects are subjugated to the rules of the game. Stories too can be regarded as internalized play, often eliciting frightening and unpleasant events but still extensions of reality, supplying strong motivation. One manifestation of this is our near-universal predilection, at least in modern industrial societies, for crime fiction, which I discuss later. First, though, let's examine something of the history and universal themes of stories through the ages—the stories that effectively define our cultures and the ways in which they operate.

< 105 >

Old Stories

The earliest stories may well have been told gesturally. Of course we have no direct record of this, although mime artists go back to ancient Greece and Rome and persist with performers such as Marcel Marceau. Dance can also tell stories wordlessly, especially in the form of ballet, which originated in the Italian Renaissance of the fifteenth century. In classical Indian musical theater, the performer tells stories using stylized gestures and hand positions to depict characters, actions, and landscapes. Another complex Indian form is Kathakali, in which Indian epics are played out with hand gestures and body movements, accompanied by songs and drumming. Perhaps closer to the linguistic telling of stories are those told in sign language by deaf communities—you can easily download examples from the Internet. I can't understand them, as sign languages are true languages, as opaque to those who don't know them as are any other foreign languages, but it is impossible to miss the vivacity with which they are told.

Aside from mime and dance, stories told through the medium of spoken language are ubiquitous and probably have a long history. Virtually all hunter-gatherer societies tell stories, usually around the fire at night, which suggests that storytelling goes back well before the arrival of literacy. The Ju/'hoansi Bushmen of southern Africa talk during the day about economics and gossip, but at night they gather round the fire and tell stories, some of which go on for days,[10] presumably being resumed each night after the day's activities, much as we television addicts follow soap operas. Nighttime stories often invoke gods, legends, and magic, along with singing, dancing, and the exchange of gifts. This division between what is spoken of during the day and the tales told at night seems to be universal. The Ainu hunter-gatherers of Japan have a saying that "the daytime is for activities, the night for deities and demons."[11]

Fire itself may have been critical. The archaeological record suggests that our prehuman ancestors had some control of fire from at least a million years ago[12] and used fire regularly beginning about four hundred thousand years ago.[13] The control of fire had a signifi-

< 106 >

cant impact on human evolution in many ways. It provided warmth and kept predators at bay. It led to the cooking of food, which lowered the cost of foraged food and the cost of sharing it. It provided a point of communal contact and social interaction, enabling people to bond and get to know one another. It effectively extended the day by lengthening the hours of light and altered our circadian rhythms. And of course it provided the opportunity for storytelling, away from the serious business of the day, with the flickering light and warmth of the fire contrasting with the darkness beyond to create an atmosphere of safety amid potential danger. In modern-day society the fire has been largely replaced by artificial light, but it is still during the evening that we tend to immerse ourselves in stories, whether reading, telling tales in the pub, watching soaps on television, going to the movies (or downloading them), or splurging on a night at the theater.

Some stories become part of folklore and are told over and over, serving to unite people and express their hopes, fears, and metaphysical concerns. The indigenous Australians have told stories that may well go back at least fifty thousand years, when they first arrived in Australia. Their stories are based on the Dreamtime, invoking ancestral spirits who traveled across the formless land and created sacred places. Many of the traditional stories are expressed in songs and dances that traveled widely over Australia, even through different language groups. The Dreaming includes stories about creation, land, people, animals, plants, as well as laws and customs. Sadly, some native Australians believe the Dreamtime stopped when Europeans arrived, although Australia now make substantial efforts to maintain and record aboriginal culture, much of it expressed in dramatic paintings that are now an integral part of the Australian art scene.

New Zealand Maori have a much more recent history in their adopted land, having arrived in New Zealand only around 750 years ago. They tell of the demigod Maui who had magical powers and lived in a place called Hawaiki—not, as some have supposed, Hawai'i but more likely the Austral Islands or the Southern Cook Islands, or perhaps both.[14] Although the Maori people regard Hawaiki as their

< 107 >

original home, it is also associated with the underworld and death; perhaps it is where their ancestors and the spirits live on. The story of New Zealand begins when Maui, a Hawaiki fisherman, dropped his magic fishhook over the side of the boat and felt a powerful tug on the line. With the help of his brother he pulled up a large fish, which they called Te Ika a Maui (Maui's fish), which became the North Island of New Zealand. The South Island of New Zealand was Maui's boat, called Te Waka a Maui, and Stewart Island, the smaller island at the southern end of the country, was Maui's anchor, Te Punga a Maui, which held the boat steady while Maui reeled in the giant fish. Although Maui caused these events, it was the great Polynesian navigator Kupe who discovered the new land Aotearoa—the Land of the Long White Cloud, otherwise more prosaically known as New Zealand.

That snippet of Maori legend does not do justice to the richness of Maori lore, which includes accounts of the creation of the world, stories of battles, songs, poems, and prayers. For centuries this information was conveyed orally, with some assistance from carvings, weaved patterns, and tattoos. Sir George Grey, governor of New Zealand from 1845 to 1854 and again from 1861 to 1868, learned the Maori language and scoured the country to find out about Maori mythology and translate it into English. The Maori version was published in 1854 and the English version in 1855.[15]

Epic Tales

The story of Maui might be described as an epic, although that term is normally applied to texts rather than stories relayed orally. According to John Sutherland, an epic is properly a story that is heroic in tone and serves to define and cement nationhood.[16] Aside from tales told orally through the ages, the earliest epic in written form was the Epic of Gilgamesh, which goes back some four thousand years and has been pieced together from engravings on clay tablets.[17] It originated in Mesopotamia (now Iraq), the so-called Fertile Crescent, and

< 108 >

might be said to mark the move from hunter-gathering to agriculture. Gilgamesh was a Sumerian king, part god, part man, who built a magnificent city but tyrannized his people brutally. He befriended Enkidu, a wild man created by the gods to divert Gilgamesh from his brutal oppression.

The story tells how Gilgamesh and Enkidu journey to Cedar Mountain to defeat Humbaba, guardian of the mountain. They then defeat the Bull of Heaven, sent by the goddess Ishtar to exact revenge on Gilgamesh for spurning her advances. In revenge, the gods kill Enkidu. Gilgamesh is then gripped by the horror that he might also die, and he resolves to live forever. He goes on a long journey to the end of the universe, killing lions, battling strange creatures, and eventually finding his way into the underworld, discovering the ferryman of the river of the dead and the last survivor of the primordial flood. But he fails in his quest for his own immortality and eventually dies. His quest has taught him that death is inevitable and people must learn to accept it.

The desire to conquer death, or at least delay it, is a human universal and continues in modern science—and pseudoscience. We have invented numerous ways to postpone death with pills, pacemakers, injections, transplants, operations, not to mention exhortations to healthier living. Diseases that were once a death sentence are now cured or at least controlled. In England in the seventeenth century, 150 of every 1,000 newborns died in their first years of diseases like diphtheria, measles, or smallpox; in today's England, only 5 out of 1,000 die in the first year, and only 7 out of 1,000 before the age of fifteen. Some optimistic scientists suggest that by 2050 humans will have become amortal, meaning that in the absence of fatal trauma their lives could be extended indefinitely.[18]

Other epics have similarly grown out of legend and tell of great heroes and superhuman exploits, defining universal themes but also serving to establish nationhood. Homer's *Iliad* tells of the Trojan War—the clash between two emerging city-states, Troy and what will become Greece. Troy is destroyed so that Greece can arise from its ashes. The emergence of Greek civilization is further chronicled in

< 109 >

Homer's second epic, *The Odyssey*, in which the Greek hero Odysseus makes an adventure-ridden return from the Trojan War to Ithaca, eventually establishing peace and stability. Other epics that served to celebrate the emergence of great and powerful nations are Virgil's *Aeneid*, telling of Aeneas (also a character in the *Iliad*), who traveled from Troy to Rome to establish the Roman Empire, and *El Cantar de Mio Cid*, telling of the Castilian hero Mio Cid, who fought in the Reconquista, the retaking of Spain from the Moors. John Milton's epic *Paradise Lost* was written in the mid-seventeenth century, when Britain was emerging as a great power.

John Sutherland wonders whether the rise of the United States as a superpower has been celebrated in an epic—or is yet to be. Some suggest that Herman Melville's *Moby-Dick* (1851) may meet the requirement, but Sutherland looks to the movies instead, nominating D. W. Griffith's movie *Birth of a Nation* (1915)—although modern US readers are likely to be offended by its racist rendition—or perhaps the western movies starring John Wayne or Clint Eastwood as a heroic cowboy. Modern polls, he notes wryly, vote *Star Wars* as the great modern epic. Perhaps we're still waiting.

Religion

From the Maui to Gilgamesh, traditional stories tell of gods and the supernatural. They form the basis of religions, helping establish the sets of beliefs and rules that bind societies and cultures together—although at the same time often creating antagonisms between cultures. Written texts have largely replaced oral traditions, vastly extending and maintaining the power of religious ideas. The Maori people, for instance, retain their traditional stories, but with the coming of missionaries and the Bible they have largely converted to Christianity. Wikipedia, itself something of a modern Bible, lists 47 sacred texts from Asatru to Zoroastrianism, including of course the Bible and the Qu'ran, which still hold millions of people in their sway. As Brian Boyd points out, religious ideas derive their power less

< 110 >

from doctrine than from stories,[19] and religious stories are typically stories of magical deeds. Psalm 77:14 proclaims, "You are the God who performs miracles; / you display your power among the people." The four Gospels of the New Testament record thirty-seven miracles performed by Jesus, including healing the sick, turning water into wine, and walking on water, and Jesus himself was held to be the Son of God, born of a virgin mother. The Qu'ran, the holy book of Islam, was revealed to the prophet Muhammad from Allah through the archangel Gabriel and was a miracle because Muhammad himself had never read or written a book and knew nothing of past events.

Belief in miracles, important in sustaining the notion that humans are unique and superior to other beings, remains strong—as we saw in chapter 2, a miracle is still invoked to explain that most human of activities, language. But the idea of miraculous powers goes beyond language. In the last week of February 2004, *Reader's Digest* published a survey which revealed that eight out of ten Australians believe that some people possess psychic powers, and seven out of ten believe in the afterlife. A majority of people also believe that it is possible to communicate with the dead and that extraterrestrials have visited the planet. In that same week, Americans voted Australia to be the country with the best image on the planet. But we shouldn't single out the Australians, because similar statistics could no doubt be compiled from most other modern societies. One survey shows that about 90 percent of people in the United States believe in God, about 70 percent believe in heaven and the afterlife, and about 58 percent believe in hell.[20]

Beliefs in magic are powerful in that they can alleviate distress and the fear of dying and offer hope. People with terminal illness naturally hope for a miracle cure. Belief in the supernatural can also serve as a mechanism of social control, more often for the benefit of despots than of the people themselves. Fear of a wrathful God or eternal damnation can encourage conformity and reduce the risk of social chaos. As Brian Boyd puts it, "We may even see reluctance to believe as a challenge to group unity and as tantamount to treason."[21] If God doesn't punish you, the people may well do so. At the time of this

< 111 >

writing, the jihad pronounced by the extremist Islamic State of Iraq and Syria has resulted in the well-publicized beheadings of infidels, and persecutions that often lead to death have been a feature of religious movements through the ages.

Beliefs in magic go well beyond the religious, though, and often depend on stories—of miracle cures, voices from the dead, premonitions, and the like. The world is awash with diets, rituals, and potions that have no curative value apart from the fact that belief itself can be curative. For that reason, perhaps, scientific investigation may never eradicate the power of irrational belief—perhaps evidence that narrative thinking, in William James's terms, takes precedence over reason. Belief in the afterlife is especially immune; it is virtually impossible to refute, because it implies another Rubicon that no mortal has been able to cross while remaining able to tell the tale.

Belief in a heavenly afterlife can also serve as solace for the poor and downtrodden, and religious teachings often praise the poor while admonishing the rich. In the Bible, Jesus says in Matthew 19: 23–24 that "it will be hard for a rich person to enter the kingdom of heaven." In compensation, perhaps, people of higher social class have been shown to be more likely than the proletariat to indulge in unethical behavior and have a more positive attitude to greed.[22] The financial collapse of Wall Street in 2007–8 may be evidence of this— although one can argue equally that it is through unethical behavior and greed that the rich get to be rich in the first place.

At least some of our belief in gods and other magical entities may have come about through our advanced theory of mind. Our ability to attribute mindlike qualities to others is often extended to apply to other species, or even to imaginary forms, animate or inanimate. In preliterate societies it was common for people to worship animals, attributing godlike properties. Notwithstanding the evidence on theory of mind in animals, discussed in chapter 5, people typically ascribe humanlike attributes to their pets, and animals that talk and behave like humans are endemic to children's stories, as I observed in the preface to this book. For some reason, bears seem to feature prominently.[23] Teddy bears are favorite companions for children,

< 112 >

and popular children's stories of bears range from *Winnie-the-Pooh* to "Goldilocks and the Three Bears." As I write, the movie *Paddington*, featuring an eponymous bear, seems to be drawing in the kids from all over town. Bear cults feature in many religions, such as Finnish paganism or the religious ideology of the Ainu of Japan. According to Greek mythology the goddess Artemis transformed a nymph into a bear and then into the constellation Ursa Major. The presence of bear bones in caves inhabited by hominins in the Middle Paleolithic has been widely taken to imply a prehistoric bear cult, perhaps involving Neandertals as well as early humans.[24] On bad days, even the stock market can turn into a bear.

The notion of an all-powerful God who created heaven and earth and monitors the behavior of humans is perhaps the most telling example of a belief in magic and, again, can be adaptive. The idea of a single God, embraced by monotheistic religions such as Christianity and Islam, is an extreme tribute to the power of humans to attribute mindlike properties even to the invisible. As the physical anthropologist Robin Dunbar points out, the notion of a God who is kind, who watches over us, who punishes, who admits us to heaven if we are suitably virtuous, depends on the understanding that other beings—in this case a supposedly supernatural one—can have humanlike thoughts and emotions. Indeed Dunbar argues that several orders of recursive thinking may be required, because religion is a social activity, dependent on shared beliefs and knowledge of what others are thinking. The recursive loops run something like this: *I suppose that you think that I believe there are gods who intend to influence our futures because they understand our desires.*[25] This is fifth-order recursion but no less complex than some of the examples discussed in chapter 7—and no less so than "The House That Jack Built," quoted in full in chapter 1. Our runaway recursive theory of mind has much to answer for.

Although religion is often seen as opposed to science, and especially to the theory of evolution, many are coming to realize that religion itself may have a scientific explanation. Darwin himself saw little mystery in the origins of religion, although of course he was acutely

< 113 >

aware that his theory of natural selection would be widely condemned by religious authorities. In *The Descent of Man* he wrote, somewhat presciently: "As soon as the important faculties of the imagination, wonder, and curiosity, together with some power of reasoning, had become partially developed, man would naturally crave to understand what was passing around him, and would have vaguely speculated on his own existence."[26]

The evolutionary biologist David Sloan Wilson makes a more direct link to evolutionary theory: "Even massively fictitious beliefs can be adaptive, as long as they motivate behaviors that are adaptive in the real world."[27] The South African archaeologist David Lewis-Williams argues that it was religion that drove the Neolithic revolution from around twelve thousand years ago to seven thousand years ago, which saw the transition from hunting and gathering to the domestication of plants and animals and the formation of large, sedentary settlements. Religion included the sacrifice of animals and humans, in part as a device to ensure that power remained in the hands of the ruling elites.[28] In that sense, perhaps, religion is a cultural adaptation to preserve peace and social order among increasingly populous settlements of people.

Perhaps it is ever so. Religion forms a large part in the rise of civilizations and in creating activities that bind people together. Besides stories, these include art, music, and architecture, especially in the form of elaborate places of worship. Science itself may derive from a religious quest to understand the world and where we came from. But there is a dark side to religion. Like language itself, it binds people together, but it also excludes outsiders. This inevitably leads to conflict, and often bloodshed, between peoples of differing religions. In his 2007 polemic *God Is Not Great: How Religion Poisons Everything*, the late Christopher Hitchens inveighed against the destruction that religion has heaped on the world and its people. Others, such as Richard Dawkins in *The God Delusion*, have focused on the irrationality of religious belief, seen to stand in the way of scientific progress.

Science itself, though, can be regarded as a story, told through countless journals and books, not to mention the Internet. Albert Ein-

< 114 >

stein, perhaps the greatest of all scientific thinkers, was well aware of the relation between science and religion:

> I maintain that cosmic religious feeling is the strongest and noblest incitement to scientific research. Only those who recognize the immense efforts, and above all, the devotion which pioneer work in theoretical science demands, can grasp the emotion out of which all such work, remote as it is from the immediate realities of life, can issue. . . . You will hardly find one among the profounder sort of scientific minds without a peculiar religious feeling of his own.[29]

We scientists seek scientific texts as avidly as ancient scholars sought scriptural ones, and we are just as beholden to cite our sources. And miracles are by no means absent from science, as we saw in chapter 2. Quarks, bosons, and leptons have some of the trappings of magic—they are invisible entities going about their mysterious business in mysterious ways. The recently identified Higgs boson has been dubbed "the God particle." Perhaps fairy tales and magic are adaptive for having prepared us for scientific discovery, and perhaps science itself is in effect a progressive refinement of our myths and legends. Ever since Darwin, it has seemed natural to consider science as in opposition to religion, but it is perhaps really part of the religious quest—part of a long process of refining ideas in line with the accumulation of knowledge about the world.

Crime Stories

In modern industrial societies, many of the themes and fantasies of traditional tales are perpetrated in our near-universal predilection for crime stories. Although murder has long featured in stories from Gilgamesh to Shakespeare's Macbeth to T. S. Eliot's *Murder in the Cathedral*, murder mysteries as a specialized form of storytelling emerged with the rise of large industrial cities, the modern equivalents of the jungle with their newfound dangers and uncertainties.

< 115 >

Murder mysteries are morality tales, because the murderer nearly always gets caught and the detective is a special form of modern hero. But there may be a dark side as well: by exposing the mistakes that lead to the criminal's being caught, crime stories may help the readers themselves avoid such mistakes and get away with murder.

The development of the crime story seems to have been ignited by a series of real-life murders committed in 1811 by one John Williams in the East End of London. These murders not only created a widespread atmosphere of fear but also led to much speculation as to the culprit. Spurred by these grisly events, the writer Thomas de Quincy wrote an essay titled "Murder Considered as One of the Fine Arts," first published in 1827 in *Blackwood's Magazine*, a British literary magazine that appeared between 1817 and 1980.

Although de Quincy's article was intended as satire, murder mysteries thereafter emerged as a literary form. An early example is Charles Dickens's *Bleak House*, published in a series of installments between 1852 and 1853—a format now taken up in modern television series that leave audiences in suspense until the murderer is revealed. Dickens's friend Wilkie Collins also published tales of murder and crime, including *The Moonstone* (1868), described by Eliot as "the first, the longest, and the best of modern English detective novels."[30] Collins also cowrote a number of short stories with Dickens. An earlier exponent, though, was the American writer Edgar Allan Poe, whose fictional detective C. August Dupin first appeared in *The Murders in the Rue Morgue*, published in 1841. Dupin had considerable influence over other fictional detectives, including Agatha Christie's Hercule Poirot and Sir Arthur Conan Doyle's Sherlock Holmes. In *A Study in Scarlet*, the first Sherlock Holmes novel, published in 1887, Dr. Watson compares Holmes to Dupin, but Holmes replies, "No doubt you think you are complimenting me . . . In my opinion, Dupin was a very inferior fellow." Conan Doyle himself was more charitable: "Each [of Poe's detective stories] is a root from which a whole literature has developed," he wrote. "Where was the detective story until Poe breathed the breath of life into it?"[31]

Crime stories are also exercises in problem solving, as the detec-

< 116 >

tive matches wits with the elusive criminal. Sherlock Holmes is the archetype here, always alert to clues as to who did it and how. Holmes became so much part of everyday people's consciousness that he was treated like a real person. When Conan Doyle killed him off in *The Final Problem*, published in 1893, public pressure was such that Doyle brought him back to life in *The Adventure of the Empty House*, set in 1894 but not published until 1903. Detective stories may hone the scientific mind; catching the criminal is a bit like catching the Higgs boson (although less expensive).

Detectives such as Holmes, along with fictional secret agents, also serve as heroes to be emulated. John Buchan's Richard Hannay, hero of *The Thirty-Nine Steps* and other spy stories, is perhaps the archetype of the values of the English public school, with stiff-upper-lipped courage honed on the rugby field—although Buchan himself was a Scot. In somewhat the same tradition are Sapper's Bulldog Drummond and Ian Fleming's James Bond, aka 007, who continues to create mayhem in popular movies. But perhaps the public grew weary of jingoism, and fictional detectives have generally taken on more gentle and often eccentric attributes. Agatha Christie's two famous detectives are the fastidious Belgian Hercule Poirot and the elderly spinster Miss Jane Marple, who combines knitting with crucial observation of giveaway clues. More recent examples include Henning Mankell's morose detective Kurt Wallander, Ian Rankin's dissolute John Rebus, and Sara Paretsky's tough-minded female detective V. I. Warshawski.[32] Lee Child's mysterious Jack Reacher, a man of no fixed abode, rivals Sherlock Holmes in terms of observation and deduction but adds a physical strength that moves him closer to the all-powerful gods of earlier times.

The Scottish writer J. K. Rowling, author of the Harry Potter series of fantasy stories for children, has recently morphed into a writer of crime stories for adults under the pseudonym Robert Galbraith. The crime-solving sleuth of these stories bears the rather ominous name Cormoran Strike; formerly a plainclothes investigator in the Royal Military Police, he is a rough but erudite character who had lost one leg during his earlier military career. And then there's Pre-

< 117 >

cious Ramotswe, founder of the No. 1 Ladies' Detective Agency in Botswana, which features in the novels of the Scottish author Alexander McCall Smith. So varied are modern-day fictional detectives that it must be difficult to invent new ones with distinctive characteristics. By the same token, though, a diet of detective fiction takes us into a wide miscellany of possible minds.

Fictional detectives bring out something of the voyeur in us, providing access not only to their own minds but also to the minds of the other characters in the story, and so to elements of society from which we may normally be barred. Ian Rankin, the Scottish crime writer, has this to say: "A detective is the perfect character, the perfect means, of looking at society as a whole. I can't think of any other character you could use that allows you access to any area of society. . . . [The detective] allows access to the banks, the politicians, the CEOs, the people who run business, but also the dispossessed, the disenfranchised, the unemployed, the drug addicts, the prostitutes."[33]

Romance

For all the popularity of crime fiction, the most popular stories in modern times are about romance, with a readership dominated by women.[34] This genre began with Samuel Richardson's *Pamela, or Virtue Rewarded*, an immensely popular novel published in 1740. It is the story of a fifteen-year-old servant girl abducted and nearly raped by her master; her virtue is rewarded when he eventually marries her and introduces her to upper-class society. *Pamela* established the formula for later romantic novels, which really took off with the establishment of Mills and Boon in 1908. The formula runs as follows: Boy meets Girl in the first few pages, Girl resists but eventually succumbs and becomes irrevocably committed by the end of the novel. The feminist writer Julie Bindel puts it more succinctly: "Man chases woman, woman resists, and, finally, woman submits in a blaze of passion."[35] She goes on to describe the typical Mills and Boon novels as "rape fantasy."

< 118 >

In some Western countries, romance accounts for nearly half of all fiction sold, and some women are reported to read up to thirty titles a month. In 2008, readers of Mills and Boon novels accounted for about three-quarters of the romantic fiction market in the United Kingdom, with one copy sold on average every 6.6 seconds. Although the basic formula remains largely intact, romantic novels come in a range of subgenres—historical, science fiction, medical, Latino, paranormal, suspense. In a more permissive age it is also increasingly erotic, with undertones of bondage or gay sex.[36]

The immense popularity of romance fiction may tell of the failure of men to meet the sexual and romantic ideals of women, although feminist authors like Julie Bindel tend to see romantic novels as exploitative. The male heroes tend to be strong, brutally handsome, misogynist, and dangerous—and yet there seems to be no letup in their popularity among women readers. Susan Quilliam, a psychologist and specialist in interpersonal relations, admits to having read and enjoyed a romance novel by Georgette Heyer as a fifteen-year-old and claims to still enjoy romantic fiction, though more for the complexity of a good story than for the sexual fantasy.[37] She argues that romantic novels play a powerful role in sex education and may even help to restore sex in otherwise loving relationships when sexual desire fades. But she also has misgivings, noting that romance fiction plays a large part in the issues that women bring to the clinic or therapy room, often through failure to distinguish between reality and fantasy.

Nevertheless romance and fantasy play roles in fiction well beyond Mills and Boon, adding spice to more literary or realistic depictions of social life—as in the novels of Jane Austen, or indeed in the work of modern male authors like Jonathan Franzen and Alan Hollinghurst. Perhaps all fiction is a balance between fantasy and realism, in which one or other may be dominant. Serious fiction often carries a moral message beyond that implicit in crime fiction or tales of romance, drawing attention to corruption and other social evils. The novels of Charles Dickens not only provide a vivid depiction of nineteenth-century London but were also designed to highlight the conditions

< 119 >

of the poor and bring about social reform. But Dickens also sought to simply entertain. He pioneered the serialization of novels so that readers would eagerly await each next installment—a technique that persisted in serialized radio productions and more recently in television series. He also perfected the art of literary caricature, creating such memorable but exaggerated characters as Fagin, Uriah Heep, and Mr. Pickwick.

And of course fiction continues to go well beyond caricature to characters that transcend normal human capabilities. Children's stories, in particular, are alive with talking animals, fairies, magicians, and other supernatural beings, as is well illustrated by the extraordinary success of the Harry Potter series. Is the supernatural adaptive? Perhaps the overstretching of the imagination allows us to better understand what might be possible, although it may be more often a product of wish fulfillment. If we could fly, become immensely strong, and control events with our thoughts, we could overcome many of our problems in coping with the world. James Bond, Superman, and Harry Potter belong in a long tradition of heroes with superhuman qualities.

I have of course only scratched the surface. There are other kinds of stories, such as chick lit, comic strips, horror stories, science fiction, stories for teenagers, war stories. Computer games have expanded into complex plots over which the gamer has some control, and they appear to have captured male engagement to match the female engagement in romance fiction. The different genres define the scripts that inform our lives, and they help extend those scripts so that we are better equipped to deal with the contingencies that social life imposes on us.

The essence of stories is in the mind, our internal travelogues and fantasies, not the words themselves. But of course much of the essence of stories depends on our capacity to share them. I now turn to the question of how language itself emerged to enable us to regale each other with the narratives that fill our lives, the gossip, the flights into humor and murder and religion, and even the dull accounts of golf rounds and travels abroad.

< 120 >

PART THREE

Constructing Language

The next three chapters explain how language evolved as a communication system for sharing our thoughts and experiences. These are internal to the mind, and in most nonhuman species largely impenetrable, except in limited ways through the growing ingenuity of neuroscience or inferences drawn from behavior. But we humans have invented ways to share our thoughts. That is what language is for, at least in the sense that language is understood as a communication system rather than itself as a mode of thought. Language often involves the sharing of mental travels in time and space and into other people's minds, and indeed rides on our ability to understand and relate to what others are thinking and feeling. But these mental adventures are not themselves linguistic, although in sharing them we gain much of what it means to be human. Through language we share the stories, real or imaginary, that define our culture, religion, learning, and entertainment.

Chapter 7 develops the idea that expressive language originated not in animal calls but in bodily gesture. This notion has a long and somewhat controversial history, but in my opinion it makes better sense than supposing that language emerged simply through the modification of animal cries. Chapter 8 then considers the question of how voicing was added, and eventually dominated, giving rise to speech. Chapter 9 examines how grammar emerged to shape our manual or vocal outputs into meaningful sequences, through which we can share ideas, experiences, and stories.

Chapter 10 closes the deal.

< 121 >

7

HANDS ON TO LANGUAGE

As for the hands, without which all action would be crippled and en-
feebled, it is scarcely possible to describe the variety of their motions,
since they are almost as expressive as words. For other portions of
the body may help the speaker, whereas the hands may almost be
said to speak. Do we not use them to demand, promise, summon, dis-
miss, threaten, supplicate, express aversion or fear, question or deny?
Do we not employ them to indicate joy, sorrow, hesitation, confes-
sion, penitence, measure, quantity, number and time? Have they not
power to excite and prohibit, to express approval, wonder or shame?
Do they not take the place of adverbs and pronouns when we point
at places and things? In fact, though the peoples and nations of the
earth speak a multitude of tongues, they share in common the universal language of the hands.

Quintilian (AD 35–100), *Institutio oratoria*

Quintilian was born in Spain but sent by his father to Rome to
study rhetoric, later setting up his own school. It was no doubt
his observations of gesticulating Romans that led him to remark on
the importance of gesture in normal conversation, or even in public
speaking—watch any politician or academic holding forth, especially
when they're trying to persuade. I once had my wine knocked over at
the dinner table by a distinguished linguist vehemently denying that
gesture had anything to do with language. I was tempted to say "Point
made," but feared a gesticular escalation—and perhaps a further spill.
A good wine is too precious for that.

< 123 >

But although the Italians are probably the most articulate and eloquent of gesturers, we all do it.[1] This is true, I'm told, even of blind people talking on the phone.

The hand-waving that goes with speech may well betray the origins of language itself. This idea is not new. The eighteenth-century French philosopher Abbé Étienne Bonnot de Condillac was one who thought that language originated in bodily gesture, but he was on dangerous ground because he was an ordained priest, and the Church's view was that language was a gift from God. He therefore presented his theory as a fable.[2] He imagined two abandoned children, a boy and a girl, too young to have acquired language, who found themselves in the desert after the universal flood. At first they communicated with gestures. If the boy wanted something out of his reach, "he did not confine himself to cries or sounds only; he used some endeavors to obtain it, he moved his head, his arms, and every part of his body." From this there grew "a language which in its infancy, probably consisted only in contortions and violent agitations, being thus proportioned to the slender capacity of this young couple."

The story goes on to explain how articulated sounds came to be associated with gestures, but "the organ of speech was so inflexible that it could not articulate any other than a few simple sounds." Eventually, though, the capacity to vocalize increased and "appeared as convenient as the mode of speaking by action; they were both indiscriminately used; till at length articulate sounds became so easy, that they absolutely prevailed." Actually that says it all, and this chapter could probably stop right here.

Jean-Jacques Rousseau, Condillac's near contemporary, also noted the priority of gesture. In his 1782 *Essay on the Origin of Languages*, he wrote: "Although the language of gesture and spoken language are equally natural, still the first is easier and depends less upon convention. For more things affect our eyes than our ears. Also visual forms are more varied than sounds, and more expressive, saying more in less time."[3]

Curiously, he went on to say that "only Europeans gesticulate when speaking; one might say that all their power of speech is in

< 124 >

their arms." I don't think he was correct in linking gesture exclusively with Europeans. There are many Maori, Asian, and Pacific Island students on our campus, and I observe them cheerfully gesturing to each other as they talk. But although Rousseau seemed to regard the manual gesturing of speakers to be rather exceptional, he did recognize sign language as the equal of speech: "The mutes of great nobles understand each other, and understand everything, that is said to them by means of signs, just as one can understand anything said in discourse."

He then went on to argue that the reason speech became dominant was that it was necessary for expressing emotion: "It seems then that need dictated the first gestures, while the passions stimulated the first words." But the argument seems back-to-front. Gesture can be a punch away from physical violence, and gestures like stamping and banging things can be expressions of anger. Perhaps Rousseau had in mind that animal vocalizations are primarily emotional, and our own vocalizations include laughing, crying, or howls of rage. But human speech is generally free from such emotive outbursts. Indeed, emotion is the enemy of speech, because it is difficult, often impossible, to talk coherently while laughing or sobbing.

Another to recognize the importance of gesture was the German philosopher Friedrich Nietzsche. Aphorism 216 from his 1878 book *Human, All Too Human* reads in part as follows:

Imitation of gesture is older than language, and goes on involuntarily even now, when the language of gesture is universally suppressed, and the educated are taught to control their muscles. The imitation of gesture is so strong that we cannot watch a face in movement without the innervation of our own face (one can observe that feigned yawning will evoke natural yawning in the man who observes it). The imitated gesture led the imitator back to the sensation expressed by the gesture in the body or face of the one being imitated. This is how we learned to understand one another; this is how the child still learns to understand its mother. In general, painful sensations were probably also expressed by a gesture that in its turn

< 125 >

caused pain (for example, tearing the hair, beating the breast, violent distortion and tensing of the facial muscles). Conversely, gestures of pleasure were themselves pleasurable and were therefore easily suited to the communication of understanding (laughing as a sign of being tickled, which is pleasurable, then served to express other pleasurable sensations).

As soon as men understood each other in gesture, a symbolism of gesture could evolve. I mean, one could agree on a language of tonal signs, in such a way that at first both tone and gesture (which were joined by tone symbolically) were produced, and later only the tone.[4]

Well, that just about says it all too, and it also notes the importance of the face as part of the gesturing system—a point I will take up in the next chapter.

In 1900 Wilhelm Wundt, the founder of the first laboratory of experimental psychology at Leipzig in 1879,[5] wrote a two-volume work on speech and argued that a universal sign language was the origin of all languages.[6] He wrongly believed, though, that all deaf communities use the same system of signing and that signed languages are useful only for basic communication and can't convey abstract ideas. We now know that signed languages vary widely from community to community and can have all of the linguistic sophistication of speech. Signed languages are recognized as official languages in many countries. American Sign Language is the language of instruction at Gallaudet University in Washington, DC, where a full university curriculum is offered. We shall also see that great apes can make a better fist, so to speak, of signing than of speaking.

The British neurologist MacDonald Critchley lamented that his book *The Language of Gesture* coincided with the outbreak of World War II and was destroyed in the Blitz, so he wrote a second book called *Silent Language*, which was published in 1975. "Gesture," he wrote, "is full of eloquence to the sagacious and vigilant onlooker who, holding the key to its interpretation, knows how and what to observe."[7] Critchley was a little evasive as to whether he thought lan-

< 126 >

guage originated in manual gestures, but at one point he did suggest that gesture must have predated speech in human evolution.

Giorgio Fano, an Italian philosopher, published a book titled *Origini e natura del linguaggio*, which appeared in two parts, one in 1962 and the second in 1973. It was translated into English and published as a whole in 1992.[8] Echoing Rousseau, he argued that language must have originally been mimed but accompanied by emotional cries. He based his argument partly on writing systems, which evolved from pictures. This argument seems strained, because reading and writing emerged well after language itself, and many peoples of the world remain illiterate. Nevertheless, one might perhaps regard the cave drawings in France and northern Italy, which go back some thirty thousand years, as marking the origins of pictorial language, and perhaps those drawings themselves derived from gesture. The Greek American artist Kimon Nikolaïdes suggests that manual gesture provides a natural basis for drawing,[9] and some artists, dating back to Rembrandt, use a style of drawing known as gesture drawing as a quick preliminary to more elaborate representations.

Fano noted that mime shows were fashionable in the ancient world, with the Roman actor Roscius once challenging Cicero as to who could express himself better: Roscius through mime or Cicero through speech. He doesn't say who won. He also included a chapter on gesture and imitation in anthropoid apes, but he was writing before chimpanzees were shown to be capable of learning a form of gestural language.

Fano's book in many respects echoes contemporary discussion on the gestural origins of language, but his book was seemingly unknown outside of the Italian-speaking world and seemingly largely neglected even after translation. It was instead the anthropologist Gordon W. Hewes, in an article published in 1973, who set the stage for wider discussion.[10] His argument rested partly on the discovery that great apes cannot be taught to speak but are reasonably successful at learning to use signs. In humans, moreover, language and hand preference are controlled by the left side of the brain, at least in most people, suggesting a common source. And like earlier authors, such as

< 127 >

Rousseau and Fano, Hewes appealed to sign language as evidence that language can be accomplished by the hands, without voicing.

This point was subsequently strengthened by the work of Ursula Bellugi and Edward S. Klima revealing American Sign Language (ASL) to be a full language, affected by specific brain injury in very much the same way that spoken language is.[11] Another who appreciated that sign language is a truly grammatical language was William C. Stokoe, who taught at Gallaudet University, where all subjects, including poetry, are taught in sign. Stokoe teamed with the anthropologist David F. Armstrong and the linguist Sherman Wilcox in a book proposing that language evolved from manual gestures,[12] and he was sole author of another book on the same topic that appeared posthumously in 1999.[13] Armstrong also continued to write on the gestural origins of language.[14]

Others, including me,[15] began to pick up on the gestural theory.[16] I must say, though, that some are still resistant to the idea that signed languages are true languages. At a recent conference I presented the gestural theory, only to be told by a prominent linguist that signed languages were merely mimes invented in the late eighteenth century. Nevertheless, evidence from different sources in support of the gestural theory has accumulated over recent decades. One important development was the discovery of mirror neurons.

Mirror Neurons

In 2000 the neuroscientist Vilayanur Ramachandran famously remarked that mirror neurons would do for psychology what DNA has done for biology—a remark that is in danger of being quoted almost as often as mirror neurons themselves are invoked. Mirror neurons are in many respects victims of their own success, having been used to explain not only language but also such functions such as imitation, empathy, and mind reading.[17] Disorders such as autism and even schizophrenia have been attributed to failure of the mirror-neuron system. Not surprisingly, there has been something of a backlash

< 128 >

against mirror neurons and the all-encompassing roles they are said to play.[18] Nevertheless, mirror neurons do provide some insight into language and its evolution, even if they don't unlock all of the secrets of the brain or even of language itself.

So what are mirror neurons? They were discovered in the 1980s, when Giacomo Rizzolatti and his colleagues in Parma, Italy, were working on neurons in the frontal lobe of the monkey brain, in an area known as F5, which responded when the animal made certain movements with the hand to grasp an object such as a peanut. To their surprise, they found that the neurons in the monkey brain also responded when *the experimenter* reached to pick up a peanut. These are the neurons that have come to be known as *mirror neurons*, because observation is mirrored in action. They might also be called "monkey see, monkey do" neurons. The Italian scientists went on to identify mirror neurons elsewhere in the brain, forming an extensive *mirror system* that includes regions of the temporal and parietal lobes, as well as the frontal lobe.[19]

The areas identified as making up part of the mirror system bear a remarkable homology to Broca's area in the frontal lobe and Wernicke's area in the temporoparietal area in the human brain, identified in the nineteenth century as involved in the production and comprehension of language, respectively. In the course of evolution, then, language appears to have evolved within a system that, back in our monkey days, was specialized for grasping things with the hands. Because the language areas are usually located in the left side of the brain, it appears that the system became lopsided in the course of its evolution. The idea that the mirror system might set the stage for the evolution of language emerged in the late 1990s.[20]

It is perhaps not altogether frivolous to note that the concept of grasping underlies some of the ways in which we refer to language itself—even spoken language. Indeed the word *grasp* itself is often used to mean "understand," if you grasp my point. *Comprehend* and *apprehend* derive from Latin *prehendere*, "to grasp"; *intend*, *contend*, and *pretend* derive from Latin *tendere*, "to reach with the hand"; we may *press* a point, and *expression* and *impression* also suggest press-

< 129 >

ing. We *hold* conversations, *point* things out, *seize upon* ideas, *grope for* words—I hope you *catch* my drift. Peter Fonagy and Mary Target suggest that such examples are indeed "a residue of gestural language"[21]—although, to continue the theme, some may consider this a *stretch* too far.

It is now clear that our own brains house a mirror system. A meta-analysis of 125 studies using functional magnetic resonance imaging (fMRI) revealed fourteen clusters of neurons with mirror properties. These were located in areas homologous to those identified in the monkey brain, including the inferior parietal lobule, inferior frontal gyrus, and ventral prefrontal cortex, but also included regions in the primary visual cortex, the cerebellum, and the limbic system.[22] In another remarkable study a group of investigators recorded activity directly in over a thousand single neurons in the brains of patients about to undergo surgery.[23] A significant proportion of these neurons responded to both observation and execution of movements of the hand and face. The areas included area F5, along with other areas associated with language and gesture in the frontal and temporal lobes. And not surprisingly, the human mirror system does indeed overlap with Broca's and Wernicke's areas.

It also seems to go beyond the monkey's mirror system in respects other than those to do with language. In the monkey, mirror neurons respond only when the monkey reaches for an actual object but not if the action is mimed in the absence of an object.[24] In humans, in contrast, the mirror system responds to mimes as well as to actual grasping, a development that may have paved the way to the understanding of acts referring to objects or action that are not present[25]— the critical feature of language known as *displacement*. Even more generally, the mirror-neuron region of the premotor cortex is activated not only when people watch movements of the foot, hand, and mouth but also when they read phrases describing such movements.[26]

But could gesture really be the source of language? Monkeys are noisy, chattering creatures, which might suggest that their vocalizations are the true origins of human speech. Darwin himself, while pointing to the possible role of gestures, believed animal vocalizations

< 130 >

to be critical: "I cannot doubt," he wrote, "that language owes its origins to the imitation and modification of various natural sounds, and man's own distinctive cries, aided by signs and gestures."[27] But as authors such as Chomsky and Pinker have been at pains to point out, vocal communication in other animals bears little if any relation to human language. Vocalization is primarily emotional and involuntary, adapted for conveying fear or anger or for establishing contact, but not for such human communicative needs as asking questions, conveying information, giving orders, telling stories, gossiping.

Our evolutionary history prepared us well for bodily expression. We are descended from tree-dwelling monkeys, with limbs adapted to climbing, swinging, and manipulation. This is still evident in the shoulder joints that allow us to reach straight above the head—a feature exploited by trapeze artists and kids who climb trees or swing from the climbing frame at the local park. Or bowlers in cricket, played by flanneled chaps—and increasingly by women as well— long ago descended from the trees. Life in the trees demanded precise voluntary control over the arms, hands, and fingers, not only for moving among the branches but also for plucking fruit or catching insects. Our capacity for intricate hand movements may also derive in part from a period when our forebears foraged in water, perhaps along coastal areas or by rivers and lakes, perhaps developing techniques for opening shells, as suggested in the previous chapter. Along with highly developed three-dimensional color vision, the primate body is well served with a capacity for producing and detecting voluntary action, with at least the potential for intentional communication.

Of course we humans do have voluntary control over voicing, but that control seems to have evolved only recently. We appear to be alone among the primates in that our brains include a projection via the pyramidal tract from the motor cortex to lower vocal centers— or so it has been claimed.[28] It is this projection, it seems, that gives us the voluntary control that enables us to speak our minds. Even the chimpanzee, our closest nonhuman relative, seems to have difficulty in producing vocal sounds voluntarily. Jane Goodall, who spent many

< 131 >

years observing chimpanzees in the wild, once wrote that "the production of sound in the absence of the appropriate emotional state seems to be an almost impossible task for a chimpanzee."[29] Similarly David Premack, a psychologist who has worked extensively with chimpanzees in captivity, drily remarks that chimpanzees "lack voluntary control of their voice."[30]

Goodall also tells of a young chimpanzee who discovered a banana and wanted it for himself but had difficulty suppressing the pant hoot that would have signaled the discovery to others. So he muffled the call by placing his hand over his mouth. This anecdote nicely illustrates that the hand movement was under voluntary control but the pant hoot call wasn't. I've also noted that some of my more impetuous friends will quickly put their hands over their mouths after saying something they didn't mean to say. Most of us, though, have the capacity to hold our tongues when indiscreet thoughts occur, and to frame our language carefully with conscious control.

Goodall and Premack may exaggerate slightly, though, because chimpanzees may have some ability to influence their vocalizations voluntarily, albeit only in very limited ways. One characteristic chimpanzee call is the pant hoot mentioned above, which does seem to vary between chimp cultures, suggesting a degree of modification.[31] Some communicative sounds that chimpanzees make don't involve the voice; for instance captive chimpanzees produce nonvocal sounds such as the "raspberry," also known as the Bronx cheer, by placing the tongue between the lips and blowing. Another example is an "extended grunt." Both sounds are often used to attract human attention, seemingly with intention.[32]

In chapter 5 I noted the study by Katie Slocombe and Klaus Zuberbühler showing that the screams of a chimpanzee under attack by another chimpanzee were modified in ways that depended on the audience.[33] This not only demonstrates an awareness of what another chimpanzee may be thinking but also demonstrates a degree of control over vocalization. But of course even we humans can exert some intentional control over emotional sounds, such as suppress-

< 132 >

ing laughter or crying in contexts where they might cause censure or pain. Like the chimpanzees, my granddaughter registers more intense distress if thwarted by one of her parents when the other parent is in the vicinity.

It seems, then, that chimpanzees can exert limited intentional control over their vocalizations, although generally in the context of emotional calls or attention seeking rather than in dispassionate, languagelike ways. Moreover, control over vocalization in chimpanzees pales beside the flexibility of their manual gestures. Attempts to get great apes to talk have been notoriously unsuccessful,[34] in contrast to the moderate success in teaching them something like intentional language through manual actions, either by having them point to symbols on a keyboard[35] or by teaching a simplified form of sign language.[36]

Human speech depends not only on whether vocal sounds can be produced voluntarily but also on whether we can learn to produce new sounds. Children must learn the vast numbers of sounds and combinations of sounds that make up our various spoken vocabularies. Vocal learning is a rarity in nature, largely restricted to some birds, including parrots, hummingbirds, and songbirds, and a few mammals, including bats, cetaceans, elephants—and us. The evidence for vocal learning in nonhuman primates is controversial and limited at best; for example, Japanese macaques seem able to alter the pitch of their calls to match the calls of conspecifics, but such examples are a far cry, as it were, from the cacophony of sounds, most of them intentional and learned, that we chattering humans make. Some learned sounds made by apes, such as the "raspberry" sound, involve the lips and tongue but not the larynx. One orangutan is reported to have learned to shape her lips to imitate a novel whistling sound produced by a human.[37] In a careful review of the evidence, Christopher Petkov and Erich Jarvis do concede that some vocal learning may occur in nonhuman primates, but they conclude as follows: "We would interpret the evidence for vocal plasticity and flexibility in some non-human primates as limited-vocal learning, albeit

< 133 >

with greater flexibility via non-laryngeal than laryngeal control. But they do not have the considerable levels of laryngeal (mammalian) or syringeal (avian) control as seen in complex vocal learners."[38]

The vocal calls of primates therefore seem to provide a poor platform upon which to construct a language. They are for the most part involuntary and resistant to learning. In marked contrast, primates gesture at will with their hands and learn quite intricate skills. This is especially true of the apes, the primates closest to humans in evolutionary terms. Let's see, then, whether we can construct a basis for language from gestures rather than from vocalizations.

Pointing

One simple gesture often seen as a precursor to language, and even a substitute for it, is pointing. A recent review of pointing in infants described it as "the royal road to language."[39] It is the first sign of voluntary expression and is firmly in place before the first words are uttered. Virginia Volterra and her colleagues in Rome show that pointing at objects is well established by one year of age, while words gradually emerge through the second year to match pointing by the age of two.[40] Indeed pointing seems a necessary step precisely because it enables the child to attach names to objects and set language on that royal road.

It is often claimed that nonhuman species don't point, with the possible exception of pointer dogs bred precisely to point at the locations of birds for the benefit of aristocratic hunters. Yet some primates, at least, can easily be taught to point to get food. Helène Meunier and colleagues in France taught baboons to point to the location of a container where they saw a raisin being hidden, so that an experimenter could fetch it for them. Furthermore, the baboons, just like human children, showed a preference for the right hand when pointing, whereas there was no hand preference for either group in grasping. To the extent that pointing is a prerequisite for learning

< 134 >

language, the origins of language might go back as far as thirty million years, to the common ancestors of old world monkeys and you.[41]

Contrary to received opinion, apes also point. According to Michael Tomasello, codirector of the Max Planck Institute for Evolutionary Anthropology in Leipzig, Germany, they point primarily to attract attention, as in pointing to a piece of food that is just out of reach in the hope that a watcher will retrieve it for them.[42] They will do this only if they can see that another individual is paying attention, which suggests that the action is intentional and genuinely communicative. Tomasello suggests that chimps will point only for humans, and the failure to observe pointing among chimps in the wild has led to the belief that chimpanzees don't point at all.[43] He rather cynically suggests that they can point but don't see the point of pointing to each other because they know it won't work. To them, pointing, you might say, is pointless.

One critical question is whether chimpanzees can point to absent objects. A defining feature of language is that it allows us to refer to the nonpresent, the property known as displacement. It was once claimed that chimpanzees cannot refer to absent objects and are therefore incapable of what has been termed "true reference."[44] In one study, chimpanzees and prelinguistic human infants were taught to request desirable objects by pointing to them.[45] Both could learn to point to objects in front of them, and the twelve-month-old infants readily learned also to point to locations in which known objects were hidden, but the chimpanzees did not. More recent studies, though, indicate otherwise. Heidi Lyn and colleagues adjusted the experiment by including chimpanzees and bonobos with prior experience of pointing, and they also placed a wider distance between the containers, making their locations more distinct.[46] They found the apes well able to point to a location where food was hidden, and the apes also ceased to point when the food was removed from the locations.

Lyn and colleagues concluded that minor methodological adjustments can make a big difference, and they cite evidence for reference to displaced objects in several other species as well, including dol-

< 135 >

phins, sea lions, and parrots. In chapter 2 I referred to the remarkable ability of the dogs Rico and Chaser to fetch named objects in another location, unseen at the time of request. Even dogs, then, may be capable of displacement. We are perhaps too ready to conclude that other species differ from humans without thorough examination of the evidence.

Gesticulating Apes

But the hands of apes do much more than point. The descendants of Koko and Kanzi, captive apes well known for their linguistic exploits, if allowed a few million years to evolve, seem headed for sign language or the laptop rather than the lectern. Kanzi, a bonobo, touches symbols on a keyboard with over three hundred characters to make simple requests and has also improvised gestural signs;[47] Koko the signing gorilla is said to have a vocabulary of over a thousand signs.[48] These well-publicized examples are often criticized for being contrived, because Kanzi and Koko were raised in a human environment and learned their skills under human guidance.[49] Even in their natural environments, though, apes do gesture often and with apparent intent to communicate. Chimps manually groom each other, and it has been suggested that their grooming is the precursor to language.[50] Another rather charming languagelike gesture is the "directed scratch," whereby a chimp will scratch a part of her body where she wants to be groomed by another chimp.

Amy Pollick and Frans de Waal watched chimpanzees and bonobos communicating with each other and observed their manual and bodily gestural communications to be more variable and less tied to context than their communications through voice or facial movements, as well as being more variable from one group to another. (The comparison is weakened because facial movements are themselves bodily gestures and should perhaps have been separated from vocalizations.) Variability and freedom from context are hallmarks of language. The bonobos showed more flexibility in gesturing than

< 136 >

did the chimpanzees, and only in bonobos did the combining of facial and vocal signals with manual gestures add to the impact on the recipient.[51] Perhaps bonobos, then, are on a path from manual to vocal communication, but via manual communication rather than voicing—and if we and they stick around long enough, we may be able to have a conversation with them. Sadly, though, it appears that bonobos are headed toward extinction more rapidly than we are.

In a heroic study, Catherine Hobaiter and Richard Byrne recount 266 days recording gestures made by chimpanzees in the Budongo National Park in Uganda.[52] They defined gestures as movements of the body, limbs, and head but not the face, and selected gestures that were intentional in that they were directed toward another chimp with the apparent aim of influencing that chimp's behavior. They also restricted their observations to gestures described as "mechanically ineffective," presumably to exclude acts like fighting, eating, or manipulation of objects. Most of the gestures they did record arose in the context of play.

Our intrepid pair recorded a total of 4,397 gestures involving at least 66 different kinds. They included actions like a directed push, a handshake, an embrace, and sexual display, as well as the "directed scratch," described above. The gestures could be assigned different "meanings" in terms of their effects on the audience.[53] As with language, the meanings were often loose or ambiguous. For instance, nine different gestures seemed to carry the meaning "stop that," while sometimes a single gesture had several meanings; for instance, shaking an object could mean "sexual attention," "follow me," "travel with me," or "move away." Some gestures, though, were tightly linked to a single meaning; for instance, leaf clipping directed to a female apparently invited copulation, the dutiful female responding accordingly. This engaging signal seems to have dropped away in our own species—except in the case of bus conductors whose clippings seem innocent of sexual intent.

Some actions carried out by apes seem to be mimes rather than requests. For example, one chimpanzee in the wild observed her daughter having trouble using a stone to crack a nut and then acted

< 137 >

out how to properly hold the stone.[54] Koko, a gorilla, mimed rolling a ball of clay between her hands, evidently to express "clay."[55] Among orangutans in a forest-living enclave in Indonesia, Anne Russon and Kristin Andrews identified eighteen different mimes, fourteen addressed to humans and four to fellow orangutans. These included mimed offers of fruit and other objects, enacting a haircut, requests to have the stomach scratched by scratching their own stomach and then offering the stick to the prospective scratcher.[56] Seven of the mimes were deceptive, including attempts to distract another animal to enable an act of theft. Deception is also one of the useful functions of language, well known to politicians and salespeople.

Other studies show that gestures of gorillas,[57] chimpanzees,[58] and bonobos[59] are also modified by social contexts and are sensitive to whether the recipient is paying attention or not—both characteristics of language. Again, this contrasts with the calls made by primates, which seem unrelated to whether anyone is listening or not. Chimpanzees also make and use tools, including sticks for fishing termites out of holes[60] and spears for jabbing into the hollow trunks of trees to extract bush babies, which they then eat.[61] Chimpanzees in the Laongo National Park in Gabon use tool sets comprising up to five different stick and bark tools to extract honey from hives.[62] It has even been suggested that some aspects of language, including grammatical structure, may derive in part from tool manufacture. This is discussed further in chapter 9.

The intentional nature of ape gestures, their ambiguity and flexibility, their sensitivity to context, and their sometimes mimelike nature do seem like precursors to language. Michael Tomasello even refers to them as "the original font from which the richness and complexities of human communication and language have flowed."[63] But we still have some way to go; several elements of language are still missing, or at least scarce. One is the sharing of information. Studies of the use of signs by chimpanzees[64] and bonobos[65] in their interactions with humans have shown that 96–98 percent of their signs are imperative—requests rather than statements. The remaining 2–4 percent serve no apparent function, except perhaps one of greeting

< 138 >

or scratching an itch. David Leavens and Timothy Racine take some issue with these claims, suggesting that chimpanzees do occasionally point to initiate joint attention, as when a bonobo pointed to hidden human observers while looking back at another ape or when a home-raised chimpanzee pointed to her nose when asked where her nose was. They note, though, that such pointing is not nearly as frequent as it is in one-year-old children. There may also be an element of sharing in mime, as described above, but these too seem to be rare.[66]

We humans do seem much more ready to share. Human language includes many more declarative statements than imperative ones, especially if one excludes the military. We talk to share information rather than merely request something for ourselves (although we do that too). Even pointing is often an act of sharing. This seems to be true even in one-year-old human infants, who sometimes point to objects that an adult is already looking at, showing that they understand that attention to the object is shared. Tomasello gives a number of other examples where the intention is to share rather than to receive gratification. A thirteen-month-old child watches as her father arranges the Christmas tree. Her grandfather comes into the room, and the child points to the tree for him, as if to say, "Look at the tree, isn't it great?" At thirteen and a half months, while her mother is looking for a missing refrigerator magnet, a child points to a basket of fruit, under which the magnet is hidden. Such gestures form the basis of language in that they are designed to share information. They also demonstrate that in development as in evolution, language begins with manual gestures.

Of course if chimpanzees don't point in order to share information, or only seldom do so, it may simply be that they are unwilling to share with humans. We humans are also sometimes unwilling to share our thoughts or secrets even with each other, and even in the age of Twitter. But there are further ingredients of language that are perhaps more obviously missing, even in apes that appear to have close rapport with humans. One is grammar, or the "unbounded Merge" of Chomskian theory. The so-called linguistic apes—Kanzi, Koko, and others—do show limited ability to combine symbols or

< 139 >

gestures to make requests, but this is far from the complex, recursive nature of human language, whether spoken or signed.

Beyond Apes

If we go back in time to the point, around six or seven million years ago, when our forebears, the hominins, began to diverge from the lineage that led to modern chimpanzees and bonobos, we find an ape already well adapted to making voluntary movements with the limbs but ill equipped either to make voluntary sounds or to modify them. If one were to begin to design a language system for such an animal, one would surely start with movements of the hands. This is exactly what happened when scientists started to find some success in building a languagelike system in present-day apes. When the bonobo Kanzi communicates his requests, he does so by pointing and gesturing, not by speaking. And as we have seen, apes may be capable of simple miming.

The hominin line differed from the line leading to modern apes in one respect that would surely have enhanced manual communication. The hominins were bipedal, standing and walking upright on two legs, which freed the arms and hands from any direct involvement in locomotion. Exactly why the hominins became bipedal is uncertain. It was in many respects a bad move, so to speak. The transition from being mainly horizontal to being vertical caused new stress to nearly every bone in the body. In chimpanzees the spine is comparatively stiff and gently curved, whereas in humans it is highly flexible with an S-shaped curve, making us especially susceptible to osteoarthritis and lower back pain. The weight of the body is carried by two feet instead of four, leading to all kinds of potential problems, including bunions, flat feet, hammertoes—even varicose veins.[67] We are something of a walking disaster. One is tempted to side with the pigs in George Orwell's *Animal Farm*: "Four legs good, two legs bad."

The benefits of bipedalism must have been considerable to outweigh these impediments and help take us to a less porcine mode of

< 140 >

existence—although not necessarily a happier one. There is little evidence that bipedal walking is more efficient than the knuckle walking adopted by the great apes. Chimps can move faster than even human athletes, although we humans may be better adapted to running long distances.[68] Perhaps bipedalism emerged because our forebears the hominins were increasingly exposed to open terrain rather than the safety of the forest, and upright walking gave a better sense of the spatial surroundings—although an alternative view is that bipedalism was an adaptation to wading and foraging in water.[69] However it came about, bipedalism also freed the hands to enable our forebears to carry objects, such as foodstuffs, tools, or their infants, from one location to another. But it may not be too far-fetched to suppose that the critical advantage arose precisely because the freeing of the hands provided for more effective gestural communication.

Just when bipedalism emerged is not altogether clear. Great apes are quadrupeds, walking with the knuckles of the forearms touching the ground. Until recently it was thought that our ape forebears must all have been knuckle walkers, but the recent discovery of a near-complete fossil called *Ardipithecus ramidus*, popularly known as Ardi, challenges this idea. Ardi dates from some five million years ago, close to the point at which our forebears diverged from the line leading to modern chimpanzees and bonobos. She appears not to have been a knuckle-walker but was adapted to an upright stance. Her foot was nevertheless shaped for grasping, suggesting that she was still primarily a tree-dweller.[70] Some have concluded that bipedalism, at least as a posture when maneuvering in the forest canopy,[71] if not for walking itself, may actually date from tens of millions of years ago and that knuckle-walking in gorillas and chimpanzees came about after the split from the human line. The knuckle-walking great apes may therefore have effectively closed off the opportunity to further exploit the hands for communication.

In any event, hands and arms are tailor-made for transmitting information about objects and events in the world. The hands can assume, at least approximately, the shapes of objects or animals, and the fingers can mimic the movement of legs and arms. The move-

< 141 >

ments of the hands can also mimic the movement of objects through space, and facial expressions can convey something of the emotions of events being described. The miming of events persists in dance, ballet, and the performances of mime artists such as Marcel Marceau, and we all resort to mime when trying to communicate with people who speak a language different from our own. Once in Russia I was able to successfully request a bottle opener by miming the action of opening a beer bottle, an action at first greeted with deep suspicion by the people at the hotel desk, and then with vast amusement as they understood my predicament.

In these respects gestures are *iconic*—that is, they are symbols with a perceived resemblance to what they refer to. The freeing of the hands with the advent of bipedalism no doubt enhanced the ability to construct iconic representations, enabling our forebears to communicate more effectively about states of the world. Even among our forebears, possibilities for communication introduced by bipedalism may have been gradual. As we have seen, the early hominins were still partly adapted to life in the trees and walked only clumsily on two legs. This is known as *facultative bipedalism*—an option rather than a necessity. Merlin Donald, in his 1991 book *Origins of the Human Mind*, suggested that what he called "mimetic culture" did not evolve until the emergence of our ancestor *Homo ergaster* around two million years ago. In *ergaster* and the later members of the genus *Homo*, moreover, bipedalism shifted from facultative to obligate—that is, it became obligatory—and a more free-striding gait developed. More critically, perhaps, brain size began to increase dramatically with the emergence of the genus *Homo*, as we saw earlier, which might be taken as evidence of selection for more complex communication.

As I suggested in the previous chapter, it was probably through mime that storytelling itself emerged, perhaps initially in the recounting, or replaying, of events that occurred in the course of a hunting expedition or in the relaying of plans for a future expedition. Glen McBride sets a likely scene. An alpha male has just returned to his group after a spectacular hunt. He gives a signal to stop—the rest of the band are physically present, the hunters themselves, the

< 142 >

weapons, the prey, so there is no need for words. Having taken central stage, "Alpha mimed the hunt story; he directed his hunters to tasks in their story, bringing them in. His watching troop had all observed young males playing hunting. Perhaps the mime included some abbreviating gestures like starting with Go Stream, or Antelope Baby. The mime was easy to understand; the prey was there, dead. The feast would follow."[72] McBride suggests that this requires no language, since it is set in the present and all the ingredients are understood from their physical presence.

Sign Languages

The natural successor to mime is sign language. Sign languages of course have been evident only in modern times, and indeed some of them have emerged within living memory. Even so, the study of signed languages and their emergence may well tell us about the early progression from mime to language, although in evolutionary terms the transition was probably a question of the one blending onto the other. Mimed action seems to occupy some of the same brain areas that language does. Essentially the same brain activity is elicited when (1) people watch video clips of a person performing mimes of simple actions such as threading a needle, (2) people act out what are called *emblems*, such as lifting a finger to the lips to indicate quiet, and (3) they give spoken descriptions of these actions. All three activities elicited activity in the left side of the brain in frontal and posterior areas—including Broca's and Wernicke's areas—that have been identified since the nineteenth century as the core of the language system. The authors of this study conclude that these areas have to do not just with language but with the more general linking of symbols to meaning, whether the symbols are words, gestures, images, sounds, or objects.[73] This system, in turn, may have evolved from the primate mirror system.

The use of signed language by the profoundly deaf activates the same brain areas as are activated by speech in people with normal

< 143 >

speech and hearing, and indeed modern sign languages still retain a mimetic presence. It has been estimated, for example, that in Italian Sign Language some 50 percent of the hand signs and 67 percent of the bodily locations of signs stem from iconic representations, where there is a mapping in space and time between the sign and its meaning.[74] In American Sign Language, too, some signs are arbitrary, but many more are iconic. For example, the sign for *erase* resembles the action of erasing a blackboard, and the sign for *play piano* mimics the action of actually playing a piano.[75] But signs need not be transparently iconic, and even if they are iconic, signers distinguish them from free mime; when people suffer from brain injury that affects their signing, their ability to mime is unaffected.[76] The meanings of even iconic symbols often cannot be guessed by naive observers or even by those using a different sign language;[77] concepts can often be represented iconically in very different ways.

Signed languages are nevertheless intrinsically more graphic than speech in depicting the world. The hands and arms can mimic the shapes of real-world objects and actions, and to some, signing can convey information in parallel instead of being forced into rigid temporal sequence. One can mimic the shape of a bird, say, while at the same time indicating its movement through the air. Even so, conventionalization allows signs to be simplified and speeded up to the point that they may lose most or all of their iconic aspect. For example, in American Sign Language the sign for *home* was once a combination of the sign for *eat*, which is a bunched hand touching the mouth, and the sign for *sleep*, which is a flat hand on the cheek. Now it consists of two quick touches on the cheek, both with a bunched hand shape, so the original iconic components are effectively lost.[78] In his book *The Expression of the Emotions in Man and Animals*, Charles Darwin remarked "on the practice of the deaf and dumb and of savages to contract their signs as much as possible for the sake of rapidity. Hence their natural source or origin often becomes doubtful or is completely lost; as is likewise the case with articulate speech."[79] Speech itself can be regarded as the ultimate step in conventionalization.

< 144 >

In Plato's *Cratylus*, Socrates asks, "Suppose we had no voice or tongue, and wanted to communicate with one another; should we not, like the deaf and dumb, make signs with the hands and head and rest of the body?" This indicates that sign languages have been around since at least 360 BC. Nevertheless, modern signed languages have short pedigrees, arising independently among different deaf communities in a surprisingly short time. For instance, Nicaraguan Sign Language (NSL) first emerged some twenty-five years ago when a school was established for deaf children and the children themselves invented the language, which gradually assumed more grammatical structure over succeeding cohorts.[80]

One fairly long-standing example, though, is Turkish Sign Language, which has an exceptionally large proportion of arbitrary, non-iconic signs. It may go back over five hundred years. Visitors to the Ottoman court in the sixteenth century noted that mute servants, most of them deaf, were favored in the court, probably because they could not be bribed for court secrets. These servants developed a sign language, which was also acquired by many of the courtiers. A photograph published in 1917 shows two servants still using sign language.[81] It is not known for sure whether modern Turkish Sign Language is related to that of the Ottoman court, but if it is, it supports the view that the passage of time is the critical element in the loss of iconic representation.[82]

Yet the speed with which sign languages emerge somewhat contaminates the comparison between them and spoken languages, which have evolved and diverged over tens of thousands of years. Nicaraguan Sign Language blossomed from mime to language in just a few generations of schoolchildren, and a similar transformation occurred in the newly emerged Al-Sayyid Bedouin Sign Language in the Negev Desert in Israel.[83] The autonomy and rapid rise of these sign languages suggests that language itself might well have developed and persisted in a predominantly manuofacial form until quite late in hominin evolution, and persists as an accessible option to this day. Michael Tomasello writes, "It is possible that the human

< 145 >

capacity evolved quite a long way in the service of gestural communication alone, and the vocal capacity is actually a very recent overlay."[84] Hear, hear!

This means that language itself might well have emerged, replacing mime, well before it took the form of speech. It could also mean that speech as the dominant mode arose only in our species and that a manual gestural language not unlike modern sign languages may have been predominant in Neandertals and Denisovans, as indeed supposed by the novelist Jean Auel, and perhaps even in early members of *Homo sapiens*.

Sign languages continue to thrive in deaf communities, providing a form of communication no less grammatically sophisticated and articulate than speech, but of course speech has come to dominate. When this happened, and why, is the topic of the next chapter.

< 146 >

8

FINDING VOICE

He thought he saw a Rattlesnake
That questioned him in Greek:
He looked again, and found it was
The Middle of Next Week.
"The one thing I regret," he said,
"Is that it cannot speak!"

Lewis Carroll, "The Gardener's Song"[1]

Not everyone agrees that language evolved from manual gestures, and indeed it somehow seems more natural to suppose that it evolved from animal calls. After all, as I noted in the previous chapter, nonhuman animals are indeed noisy creatures, as we humans are. So why, one might ask, should we suppose that it all began with waving of the hands? Conversely, one might even wonder why it is that we talk at all rather than use our eloquent hands to communicate. It's talk that clearly dominates our discourse, except perhaps for the generation that now transmit text messages on their cell phones, reverting to gesture.

The linguist Robbins Burling wrote, "The gestural theory has one nearly fatal flaw. Its sticking point has always been the switch that would have been needed to move from a visual language to an audible one."[2] So how am I to answer?

For a start, control of hand and mouth are integrally connected, even originating from a common brain region that can be traced back to our common ancestry with fish![3] More to the point, though,

< 147 >

throughout primate evolution the hands and mouth are connected through the process of eating. People and monkeys bring food to the mouth in exquisitely coordinated fashion. Such coupling may well carry over to language. In a classic article published in 1985, David McNeill demonstrated the tight synchrony between speech and the gestures that accompany it, concluding that "gestures and speech share a computational stage."[4] And speech is itself a system of gestures, made up of movements of the lips, the velum, the larynx, and the blade, body, and root of the tongue. One might suppose, then, that the production of language shifted from one set of gestures to another. Indeed there was probably overlap; manual gestures may well have been punctuated by grunts, and vocal gestures are generally accompanied by a good deal of hand waving, especially in Italy. Of course we perceive manual gestures visually and vocal gestures through the sounds they produce, but there are reasons to believe that we even tend to understand vocal gestures *as gestures* rather than as patterns of sound.

A strong version of this idea is what has been termed the *motor theory of speech perception*, which arose from the work of the late Alvin Liberman and others at the Haskins Laboratories in the United States.[5] They started out in the 1950s by trying to find characteristics of the sound waves that define the basic elements of speech, hoping to be able to devise an instrument that would automatically decode speech. The idea was that you could simply talk to this instrument and it would pick out these elements and then print out what you say. This would eliminate the need for typists or even keyboards.

The hopes of producing such a device were soon dashed. I am still using a keyboard as I compose these very words (although typists have largely vanished from my workplace).[6] Even the individual speech sounds, at the level roughly represented by letters of the alphabet, don't correspond to simple packets of sound,[7] so the task of inventing a device that would simply pick up these sounds and transmit them to a keyboard has proved a highly daunting one. For example, the *b* sounds in words like *battle, bottle, beer, bug, rabbit, Llareggub,* and *flibbertigibbet* probably sound much the same to you, but

< 148 >

the actual acoustic components corresponding to the *b* vary widely, to the point that they actually have virtually nothing in common. The same lack of constancy is true of other speech sounds, especially the so-called plosive sounds *b, d, g, p, t,* and *k,* with the acoustic signals varying widely depending on where they occur—*dog* and *god* don't simply involve a juxtaposition of actual sounds, or a juxtaposition of anything else, for that matter.

Liberman and colleagues concluded that we hear speech sounds in terms of how they are produced, not in terms of the sound patterns themselves. There is much more in common in the ways we produce the *b* sound in *bunny* and *rabbit* than there is in the sounds actually produced, and the *b* sound in a word like *cab* when at the end of a sentence, as in "Call a cab," scarcely exists as an identifiable sound at all. This of course greatly complicates the problem of mapping from sound to print, and machines that do this require very complex programming and still aren't much good at it.

It must be said, though, that some progress has been made through the use of computers with huge processing capacity and algorithms that learn to recognize patterns. One approach uses so-called deep neural networks, with an input layer that receives acoustic information in binary form, and many "hidden layers" with interconnections that adapt to lead to desired output. One approach has been to "train" such systems to decode broadcasts of the BBC News. In one report, the most successful system still wrongly identified 17.5 percent of the words after fifty hours of training.[8] Such approaches rely on the acoustic input, and they may indeed continue to improve, but it is difficult to avoid the impression that reference to how speech sounds are actually produced would help—or perhaps those hidden layers will eventually discover this through their adaptive adjustments. But for now, the only instrument that can decode speech with facility remains the human brain.

This idea that speech sounds are heard in terms of how they are produced brings to mind the mirror-neuron system discussed in the previous chapter, whereby how you see an action is mapped onto how you produce it. This system seems to work for action-produced

< 149 >

sounds as well as for visual inputs—even in monkeys. For example, some mirror neurons in the monkey brain respond to the sounds of actions, such as the tearing of paper or the cracking of nuts.[9] But in monkeys, unlike humans, mirror neurons are deaf to vocalizations—they don't respond to the sounds of monkey calls. Only later in primate evolution did the mirror system encompass vocal sounds themselves, enabling us to perceive speech in terms of the way it is produced rather than in terms of how it actually sounds. Among the primates, it may have been only we humans who gained the benefit of incorporating of speech sounds into the mirror system—but more of that later.

Language also depends on memory—our ability to remember symbols, be they gestures, spoken words, or printed ones. Compared to vision, sound appears to provide a poor basis for memory. Monkeys are much better at remembering visual patterns than auditory ones[10]—another strike against the idea that language evolved from primate calls. We humans, too, have difficulty remembering patterns of sound unless they are related to gestures or have meanings. One study shows people to be much better at remembering spoken words, pronounceable nonwords like *glaik* or *breet*, or sounds that are tied to actions (such as a bell or a xylophone) than at remembering words played backwards, even only five minutes after hearing these sounds. The authors of this study suggest that their findings are evidence for a motor theory of auditory memory. They conclude that "speech and auditory memory are so critically dependent on each other that they had to co-evolve."[11]

But the motor theory can't be the whole story. We saw in chapter 2 that apes, and even dogs, can learn to understand spoken words, at least to the point of being able to respond to spoken commands or pick out objects designated by their names. Chaser the border collie has a receptive vocabulary of 1,022 spoken words. Dogs have been selectively bred by humans, so they are perhaps exceptional, but even chinchillas and quails appear able to discriminate human speech sounds in much the same way that we do. Yet none of these animals can actually articulate words, so the motor theory doesn't

< 150 >

apply to them. People with so-called Broca's aphasia, resulting from damage to the frontal-lobe areas responsible for producing articulate speech, can understand speech even though they can't produce it. Even babies can understand words before they are able to articulate them. Indeed a recent study suggests that babies can encode and remember the sounds of words within the first four days following birth![12]

It's fair to conclude, then, that both the way we produce speech and the auditory input itself contribute to the way we hear speech.[13] Gregory Hickok and David Poeppel have suggested what they call a "dual stream" theory of speech perception.[14] One stream is lower in the brain: the so-called *ventral stream*, which progresses from the temporal lobe to the frontal lobe of the brain. This stream is responsible for the understanding of speech, and it is presumably this stream, or its homologue, that is responsible for the limited understanding of human speech in Kanzi or in dogs like Rico and Chaser—and perhaps even in your pet cat, not to mention inarticulate teenagers. The other stream is higher: the so-called *dorsal stream*, which flows from the parietal lobe and is responsible for the actual production of articulate speech. It is this component that is lacking in dogs and apes.

Yet the distinction between dorsal and ventral systems actually originates from studies of monkeys and applies to vision and manual action rather than hearing and voicing.[15] In vision, the dorsal system is specialized for translating perception into action, as in reaching for an object in space or inserting a letter into the slot in the postbox, while the ventral stream is programmed to recognize what objects are.[16] They are sometimes known as the "where" and "what" systems, respectively. What may be unique to humans, at least among the primates, is not the dorsal system itself but its adaptation for the learning and production of vocalization—although a homologous circuit, both anatomically and genetically, also seems to distinguish birds that learn songs from those that don't.[17] In the monkey brain, the dorsal circuit is specialized for controlled movements of the hands and face but not for vocalization, leading Michael Petrides and Deepak Pandya to remark that the system "has been adapted, in

< 151 >

the human brain, for use in sound-to-speech articulation transformations."[18] This of course is consistent with my view that language originated in manual and facial gestures but later expanded to include voicing.

The dorsal system is part of the mirror system and may indeed play a role in perceiving as well as in producing action, as implied by the motor theory of speech perception. In an extensive critique of mirror-neuron theory, Greg Hickok suggests that the mirror system is not so much a device for tuning into speech sounds as a "feed-forward" mechanism for calibrating the production of speech sounds.[19] We shape our own speech sounds in terms of how we hear them—this, perhaps, is why babies babble before they learn to speak. They're calibrating. The ability to speak may well modify the way we perceive speech, but as the examples of dogs and apes show, perception of speech can occur without any influence of speech production. Hickok gives the analogy of observing a person playing the saxophone. A non–saxophone player may well have adequate perception of the player's performance, but a fellow saxophonist is likely to have an enriched perception and understanding by relating her perception to her own experience.

The Myth of Arbitrariness

Although speech itself can be regarded as gesture, it seems to differ in another way. Whereas gesture can provide iconic representations of real-world events, speech seems to be made up of sounds that seldom bear any relation to what they represent. Indeed the Swiss linguist Ferdinand de Saussure wrote of the "arbitrariness of the sign" as a defining property of language,[20] and on this basis it is sometimes supposed that signed languages, with their strong basis in iconic representations, are not true languages.

Even spoken words, though, are not entirely arbitrary. Words sometimes do reflect the shapes of things they name—that is, there is an iconic component to speech as well as to manual gesture. The

< 152 >

German psychologist Wolfgang Kohler showed people drawings of two meaningless objects, one smooth and rounded and the other sharply inflected and spiky, and asked people to choose which one was named "baluma" and which "takete." Ninety-five percent of respondents chose "baluma" for the rounded one and "takete" for the spiky one.[21] This has been repeated many times with slightly different names. Vilayanur Ramachandran and Edward Hubbard attribute this to synesthesia—a natural (though widely varying) tendency to associate stimuli of different modalities, in this case sound and vision.[22] They also suggest more direct mappings. For instance, the words in various languages for referring to someone in the second person (*you* in English, *tu* or *vous* in French, *thoo* in Tamil) involve pushing the lips forward to the listener, while words referring to the self (*me* in English, *moi* in French, *naan* in Tamil) seem to point inward with tongue and lips to the speaker herself.

It has also long been observed that some sounds, such as the /i/ in *little*, suggest smallness, and others, such as the /a/ as in *large*, suggest largeness.[23] Another observation is that cross-linguistically words for "nose" contain a higher than average proportion of nasals, while words for "lip" have more than average numbers of bilabials.[24] Speech can also mimic visual properties in subtle ways; for example, it has been shown that speakers tend to raise the pitch of their voice when describing an object moving upward and lower it in describing a downward movement.[25] Even speed matters: People respond to pictures more quickly if a spoken sentence describing the picture matches the motion represented in the picture. They match a moving object, such as a galloping horse, more quickly to the sentence if the sentence is spoken quickly, and match a stationary object more quickly if the sentence is spoken relatively slowly.[26] Even three-month-old babies associate pitch with height and thickness, with sounds of higher pitch linked to increased physical height and narrower shapes, associations that may govern later development of language.[27]

And of course there are words that are onomatopoeic. One such word is *zanzara*, which is the evocative Italian word for mosquito, and Steven Pinker notes a number of newly minted examples: *oink*,

< 153 >

tinkle, barf, conk, woofer, tweeter[28]—although the last-named now seems to mean something else.

Ramachandran and Hubbard go so far as to suggest that synesthesia solves "the riddle of language origins,"[29] evidently identifying language with speech. Clearly, though, manual gesture offers much more flexibility and mimicry for the descriptive labeling of objects, especially in a world where vision dominates. Sir Richard Paget, an English baronet, argued in 1928 that sound symbolism might itself be based on manual gesture. His own observations of mouth movements during speech led him to the conclusion that the movements of the tongue and lips might "copy the pantomimic movements of other organs, such as the human hands and arms"[30]—another endorsement of the gestural theory, although it's perhaps a bit of a stretch. The distinction between gesture and speech, then, is really one of degree rather than kind, and both signed language and speech carry elements of physical description and arbitrary symbol. As explained in chapter 7, the shift from one to the other depends on *conventionalization*, the gradual tailoring of a descriptive system to one that is governed by convenience and efficiency and carried by culture rather than by the representational power of bodily movement.

Iconicity, then, continues to play a role in language, in development as much as in evolution.[31] But even within signed languages, representation become less iconic, or "pictorial," and more arbitrary over time, as we saw in the previous chapter. Speech is of course more arbitrary than signed language, and the increasing role of culture means that some six thousand languages have emerged with vastly different vocabularies, each more or less opaque to every other. Because it conveys less pictorial information, speech probably emerged later in the process of conventionalization. Perhaps early language was made up largely of mimed gesture accompanied by involuntary grunts, but our forebears gradually assumed intentional control over those grunts, which could then be included in the process of conventionalization. The grunts became refined and eventually took over, although the hands continue to provide orchestration.

< 154 >

Facing Facts

To return to Burling's sticking point, though, even the gap between manual and vocal gestures may still seem too great a Rubicon[32] for evolution to cross. But there is a natural bridge—the face. The use of facial expressions as social signals probably goes far back in primate evolution. The range of expressions increases as a function of body size, and gorillas and chimpanzees have almost as many different expressions as we humans do.[33] A classic example is the bared-teeth display, used to induce submission but also to promote social bonding. Another example is lip smacking, which is used to promote bouts of grooming. Of course many facial expressions are emotional, indicating rage, fear, or happiness, and in humans these have little to do with language—and indeed may interfere with language, as when you try to speak coherently while laughing or sobbing. According to one hypothesis, moreover, facial expressions are signals of trustworthiness or its lack, and good poker players work hard to suppress them. As the American actress Gena Rowlands said, "I can never have a poker face. Anybody looking at me can tell exactly what I'm thinking."[34]

But in primates the motor cortex also contributes to movements of the face and mouth, suggesting a degree of voluntary control.[35] To some extent, at least, primates can deliberately alter facial movements, suggesting that their facial expressions may not always be trustworthy—beware the smile on the face of the chimp, if not of the tiger. We humans, of course, can also move our faces independently of speech itself. Facial movements are an important component of sign languages,[36] and people conversing in sign language watch the face as much as they watch the hands—in some contexts more so.[37] In the course of evolution, then, voluntary communication probably shifted increasingly from the hand to the face. Movements of the head and face accompany much of our normal conversation as we nod in agreement, shake our head in disagreement, frown in disapproval, or raise our eyebrows to show surprise—or superiority. The haka, a war dance performed by the Maori people, involves fearsome movements

< 155 >

of the protruding tongue and rolling eyes, along with other bodily gestures.[38] Sticking out one's tongue and pulling faces are cheeky (*sic*) ways of expressing ourselves and are known to children all over the world. Unlike the facial expressions that accompany emotions such as laughing or crying, such untoward behavior is voluntary.

It was but a small step from the external surface of the face to the movable parts inside it. Speech is facial gesture, half swallowed. Most of the movement in speech gestures is not visible to the watcher, so access to this movement must be accomplished through the medium of sound. We therefore evolved the capacity to create sound voluntarily through movements of the vocal folds, a membrane stretched across the larynx, and this sound is modulated by the gestural movements of the tongue, velum, and lips. This modulation enables the listener to infer the gestures that produce the words. Even so, some visual components remain and can still feature in the perception of speech. Deaf people, especially, can become skilled at lipreading, and poor synchronization of sound and mouth movements, as in badly dubbed movies, can interfere with perception of what is actually said.

Indeed, people sometimes hear what they see rather than what is contained in the sound. The psychologists Harry McGurk and John MacDonald dubbed sounds, such as *ba-ba*, onto videos of a mouth that was actually saying something different, such as *pa-pa*, and the viewer/listeners often reported hearing what the speaker was seen to be saying (*pa-pa*) rather than the speech sound itself (*ba-ba*), or sometimes a blend of the two.[39] Other studies show the parts of the brain involved in producing speech are activated when people simply watch silent videos of people speaking.[40] Ventriloquists know the power of vision when they project their own voices onto the face of a dummy by synchronizing the mouth movements of the dummy with their own tight-lipped utterances.

The anthropologist Robin Dunbar has suggested that the transition from grooming to speech may have derived from laughter. As a bonding device, laughter is more effective than grooming, especially as groups grow larger and they can all have a good laugh together. Some modification to the mechanics of laughing must have occurred,

< 156 >

though, because chimpanzees laugh on both the incoming and out-going breath, whereas humans laugh only on the outgoing breath.[41] This change was perhaps a precursor to speech in that speech is pro-duced only as we breathe out. Genuine laughter, though, is involun-tary. Robert Provine, in his book *Laughter: A Scientific Investigation*, records that a girls' boarding school in Tanganyika (now Tanzania) had to be closed because of an uncontrollable epidemic of hysterical laughter. A similar episode is recounted in the Scottish poet Carol Ann Duffy's narrative poem "The Laughter of Stafford Girls' High." Duffy, who is poet laureate of the United Kingdom, attended Stafford Girls' High in Staffordshire, England.

For laughter to serve as the basis for speech, though, our laugh-ing predecessors must have gained voluntary control—laughter is notoriously hard to fake. Dunbar suggests that this transition oc-curred around when the genus *Homo* emerged, now dated to some 2.8 million years ago.[42]

Be that as it may, joke or no joke, the transition from hands to face, and in particular from hand to mouth, can be understood in biologi-cal terms. In primates, hand and mouth are closely linked both in the brain and in behavior. In the motor cortex of the brain, responsible for initiating body movements, the so-called hand area is adjacent to the mouth area. Some neurons in the frontal lobe of the monkey are activated when the animal makes a grasping movement either with the hand or with the mouth.[43] The work of Maurizio Gentilucci in Parma, Italy, shows a close connection between movements of hand and mouth in people as well. If human subjects are told to open their mouths while grasping objects with their hands, the size of the mouth opening increases with the size of the grasped object.[44] Grasping movements of the hand also affect the way we utter sounds. If people are asked to say "ba" while grasping an object, or even while watch-ing someone else grasp an object, the syllable itself is affected by the size of the object grasped. The larger the object, the wider the opening of the mouth, with consequent effects on the speech sounds them-selves.[45] Even one-year-old babies show these effects.[46] These links between hand and mouth probably originated in eating rather than

< 157 >

communicating, having to do perhaps with preparing the mouth to receive an object after the hand has grasped it, but they were adapted for gestural and finally vocal language. Again we see that evolution tinkers with what's already there—it seldom creates anew.

The gradual transition from body to face to mouth is an example of *miniaturization*—a common feature of communication systems, as evidenced by present-day cell phones and microchips. My first lab computer was the size of a home refrigerator. Speech is of course much more compact than pantomime and much more energy efficient. I have been told that instructors of sign language often need massage after an exhausting day of moving their arms and bodies. In contrast, the physiological costs of speech are so low as to be nearly unmeasurable.[47] In terms of the energy expended, speech adds little to the cost of breathing, which we must do anyway to keep alive. Some people never seem to tire of talking.

With the production of language neatly tucked away into the mouth, the rest of the body, and especially the hands, were largely freed for other activities—a second freeing, as it were, after upright walking relieved the hands from locomotory duty. So the devil again found things for idle hands to do. These no doubt included the making and use of tools and weapons, writing and drawing, and gentle evening games of tennis. These activities would have been inhibited, or perhaps would never have evolved, had we persisted with manual language, a point that did not escape the attention of Charles Darwin: "We might have used our fingers as efficient instruments, for a person with practice can report to a deaf man every word of a speech rapidly delivered at a public meeting, but the loss of our hands, while thus employed, would have been a serious inconvenience."[48] I suggest later that the switch to speech may have led to the remarkable explosion of technology in our species.

Speech is also a blessing to the blind, although the extraordinary verbal talents of Helen Keller, who contracted an illness at the age of nineteen months that left her deaf and blind, show that language can flower in the absence of both sight and sound.

For the sighted population, though, speech freed the eyes. You can

< 158 >

listen to speech with the eyes closed, or while concentrating on some manual task while someone tries to tell you what to do. This may have been important in the development of pedagogy. Unlike signing, speech enables people to communicate at night or when speaker and listener are not in visual contact. Mary Kingsley, the noted British explorer of the late nineteenth century (and niece of Charles Kingsley, whom we met in chapter 4), made the following observation of tribes she encountered in Africa: "[African languages are not elaborate enough] to enable a native to state his exact thought. Some of them are very dependent upon gesture. When I was with the Fans they frequently said 'We will go to the fire so that we can see what they say,' when any question had to be decided after dark, and the inhabitants of Fernando Po, the Bubis, are quite unable to converse with each other unless they have sufficient light to see the accompanying gestures of the conversation."[49] This may sound condescending, and I have no idea whether her observations were accurate, but Kingsley did at least recognize that signing could provide a basis for articulate language.

In outlining the advantages of speech over gestural language, I am often berated by the deaf or by those who work with them, who tell me that signed language creates no impediments—except of course a deficit in the understanding of speech, although many deaf persons become quite proficient in lipreading. Deaf signers sometimes claim a superior culture and refer to themselves as Deaf with a capital D. Deaf parents within this culture often do not wish their hearing children to learn to speak, and some prospective parents have stated their willingness to use genetic testing to ensure their children will be born deaf or to consider in vitro fertilization of a deaf embryo. Deaf people have told me they have no difficulty with other manual operations, such as driving a car, while engaged in signed conversation. And of course there are situations where gestural language undoubtedly has advantages over speech. One might suppose, for example, that early hunter-gatherers benefited from using silent forms of communication rather than letting their voices disturb prey or alert predators. To my mind, at least, the overall advantage lies with vocal over manual

< 159 >

language, and that's why we speak rather than sign—but as a speaker rather than a gesturer, I may well be biased.

My compatriot Peter MacNeilage has argued against the gestural theory of language origins on the grounds that any advantages of speech over signing seem too insubstantial to have led to the demise of early signing.[50] Well, we could still sign if we chose to—sign languages are learned as effortlessly as speech—but we don't, unless circumstances force us to. We might also remember that any advantages of speech over signing need only have been slight. The biologist J. B. S. Haldane computed that a variant resulting in a 1 percent gain in fitness would increase in frequency from 0.1 to 99.9 percent in just over four thousand generations.[51] This may suggest a conservative estimate of about one hundred thousand years for the conversion of a language that was largely manual to one that became capable of autonomous speech. Ezra Zubrow computed that a difference of only 1 percent in mortality rates between Neandertals and *Homo sapiens*, who overlapped geographically, could have led to the extinction of the Neandertals within thirty generations, well under a millennium.[52] It is conceivable that our own forebears were further advanced in the evolution of speech, which gave them the critical edge over their hapless cousins—but more of that later.

The notion that language shifted progressively from hand to face to vocal tract is of course something of a simplification. Even today, language involves all three kinds of gesture. I suggested earlier that early communication may have been pantomimic but was probably punctuated by grunts. Modern humans can get by well enough simply through the sound of the voice, but in normal conversation we also gesticulate, point, raise fingers to signify quotation, nod, raise eyebrows—a *son et lumière* performance. Still, speech itself has become the dominant mode. The gestures that go with it can certainly help get a message across and can sometimes substitute for words, as when we point rather than give verbal directions or shrug when we don't know the answer to a question. But take away the visual accompaniment and speech can carry the message, as on radio or the ubiquitous cell phone. Take away the vocal component, though,

< 160 >

and most of the message is lost—unless of course you know sign language.

Some influential authors have accepted that manual gestures are a critical component of language but have maintained that speech and gesture evolved together. Adam Kendon, a pioneer in the study of the manual gestures that habitually accompany speech, dismisses what he calls the "gesture first" theory in favor of the idea that vocalization and manual gestures were equal partners in the evolution of language.[53] David McNeill, also a groundbreaking figure in the study of gesture, argues similarly that the close synchrony between speech and gesture must mean that they evolved together. He concedes that our forebears may once have communicated with manual gestures, but he claims that if they did it could not have led to language; it would have led to pantomime, which does not synchronize with speech.[54]

The idea that synchrony dictates a joint origin is nevertheless debatable. Much of the movement of humans and animals is controlled by what has been termed the *central pattern generator*, which underlies the rhythmicity of walking or swimming, as well as such basic movements as breathing, sucking, and eating. Speech itself is a rhythmic activity, partly synchronized with breathing. Given the rhythms of the body, it would not be surprising if speech and gesture fell into synchrony, regardless of which appeared first in the evolutionary sequence. Moreover, there seems no reason to suppose that gestural communication would lead inevitably to pantomime if there proved to be advantages in adding a voiced component, since the gestures themselves would be easily entrained by the rhythmicity of articulation and the constraints imposed by breath control.

The gradual dominance of vocal gestures, leading to speech, would have imposed further constraints, to which the manual component would conform. Speech requires that communicated information be *linearized*, piped into a sequence of sounds that are necessarily limited in terms of how they can capture the spatial and physical natures of what they represent. The linguist Charles Hockett put it this way: "When a representation of some four-dimensional hunk of life has to be compressed into the single dimension of speech, most iconicity is

< 161 >

necessarily squeezed out. In one-dimensional projections, an elephant is indistinguishable from a woodshed. Speech perforce is largely arbitrary; if we speakers take pride in that, it is because in 50,000 years or so of talking we have learned to make a virtue of necessity."[55]

The question now is, when did speech take the commanding role?

Back to the Beach

In evolutionary time, it is unlikely that the emergence of speech was sudden. Rather, language was probably a combination of sight and sound, as indeed it is today, but with the vocal component gradually increasing, diminishing the role of gestures. One perspective on this process comes from the aquatic ape hypothesis (AAH), summarized in chapter 5. Mark Verhaegen argues that the aquatic phase set a platform for the evolution of speech. As we have seen, other primates have little if any voluntary control of voicing. The ability to breathe voluntarily, he suggests, was an adaptation to diving, where you need to hyperventilate before plunging and then holding your breath during the dive. The fine-motor control over lips, tongue, velum, and throat necessary for producing consonants evolved for the swallowing of soft, slippery foods such as mollusks without biting or chewing. Think of oysters and white wine.

Philip Tobias once suggested that the term *aquatic ape* should be dropped, as it had acquired some notoriety over some of its more extravagant claims. Verhaegen suggests that the aquatic theory should really be renamed "the littoral theory," because early *Homo* was not so much immersed in water as foraging on the water's edge, diving or searching in shallow water, and probably also roaming inland.

Even in its modern form the aquatic ape hypothesis (AAH) remains controversial. Verhaegen quotes *Wikipedia* as asserting that "there is no fossil evidence for the AAH"; he disagrees, citing evidence that "virtually all archaic *Homo* sites are associated with abundant edible shellfish."[56] But perhaps it is fair to conclude that our *Homo* forebears were characterized not so much by their aquatic exploits

< 162 >

or their life on the savanna as by the sheer diversity of their activities. This diversity is evident in modern human life. Trapeze artists and gymnasts remind us that we retain some heritage of our earlier life in the trees, and modern Olympics show that we are capable of extraordinary skills in swimming and diving, as well as in running, jumping and throwing things. But in modern life we have entered a phase in which beach and savanna are primarily recreational, replaced by cityscapes and dreary offices lit with computer screens (where, indeed, I find myself now).

And So to Speak

A littoral phase would have helped shape communication through manual gesture as well as supporting the voluntary control of vocalization, but there remains good reason to suppose that manual communication came first, for reasons already stated. Voluntary control of the limbs long precedes voluntary control of the voice in primate evolution, and our manipulative primate hands provide much the more natural way of representing events in the physical world. The question then is when and how vocalizations were incorporated, to eventually become dominant in the hearing population.

The transition was probably gradual, and involuntary sounds such as grunts may always have accompanied communicative gestures. Kanzi the bonobo vocalizes prolifically while communicating through gesture or by pointing to symbols on his keyboard, but this is probably largely emotional and it is the gestures that provide the information. The ease with which sign languages are acquired suggests that manual gestures may even have been dominant until quite late in *Homo* evolution. Jean Auel's novel *Clan of the Cave Bear* is set in prehistoric Europe some twenty-seven thousand years ago, when early humans and Neandertals coexisted. A five-year-old girl, Ayla, is orphaned after an earthquake kills the rest of her family, and she eventually joins a Neandertal community. The Neandertals in the story cannot speak, so they communicate in sign language. I should

< 163 >

not of course take a fictional novel to be acceptable scientific evidence, tempting as it is to do so, but Auel is something of an expert on early humans and Neandertals, and the use of sign language by Neandertals is a theme in her other novels as well. Curiously, though, the Neandertals in *Clan of the Cave Bear* not only cannot speak but also cannot laugh or cry, and when they see Ayla weeping they think she has an eye disease.

Verhaegen suggests that littoral features were more prominent in *Homo erectus* than in the Neandertals, implying that the Neandertals may have lacked preadaptations leading to speech—although one report suggests that the Neandertals ate salmon.[57] The question whether the Neandertals could speak is fraught. The linguist Philip Lieberman has long argued that the shape of their vocal tracts would have denied them articulate speech.[58] Robert McCarthy of Florida Atlantic University used reconstructions of the Neandertal vocal tract to create simulations of a Neandertal vocalization, which sound more like a sheep or a goat than a human.[59] Humans are able to speak because the larynx descended in the throat creating a right-angled tube, so that the sound waves created by the larynx could be modulated in several ways. For example, the back of the tongue can be lifted and released against the angled portion at the back of the mouth to create sounds such as the *g* sound in *goat*, and the front of the tongue can be lifted and released against the back of the teeth to create the *t* sound in *toad*. The lips come into play with *bees* and *peas*. The shaping of the vocal tract and the muscular control over it are also critical to creating the varied vowels, from *ee* to *oo*. The descent of the larynx came at a cost, because it opened up the breathing tube at the back of the mouth to the possibility of food "going down the wrong way," making humans peculiarly susceptible to choking. According to the National Safety Council, choking on food is the fourth leading cause of accidental death in the United States, around a tenth of the deaths caused by motor vehicles.[60]

Another feature of the human face is that it is relatively flat compared with those of other apes and primates. Even in the Neandertal, the face tends to protrude; compared to them we humans have a

< 164 >

shorter sphenoid, the central bone of the base of the cranium from which the face grows forward. Daniel Lieberman suggests that this flattening ensures that the right-angled vocal tract has equal horizontal and vertical components, a configuration that increases the ability to produce the full range of distinct speech sounds.[61] The human brain is also more rounded than the Neandertal brain, perhaps reflecting enlargement of temporal and frontal lobes, which may be critical to the control and processing of speech.[62]

As recently as 2007, Philip Lieberman wrote that "fully human speech anatomy first appears in the fossil record in the Upper Paleolithic (about 50,000 years ago) and is absent in both Neandertals and earlier humans."[63] This ties in remarkably with the notion of Chomsky and others, discussed in chapter 2, that language evolved as recently as fifty thousand years ago, except that Lieberman's conclusion applies to speech rather than to language itself. By now, I hope, the reader will understand that language should not be equated with speech.

But not all are content to reduce the Neandertals to silent gesturing and perhaps incoherent spluttering. The French linguist and anthropologist Louis-Jean Boë argues that Lieberman's reconstruction of the Neandertal vocal tract is incorrect and that there was probably no physical impediment to speech.[64] Lieberman is defended, though, by his one-time PhD student Tecumseh Fitch,[65] and Boë is careful to point out that the Neandertals may not have had the necessary brain organization to enable them to speak. A yet more recent reconstruction suggests that the Neandertal vocal tract did not differ substantially from our own.[66] Until a Neandertal walks through the lab door and offers herself for inspection, or simply speaks, this intriguing debate is likely to continue.

The FOXP2 Saga

The Neandertal drama has also played out in a different arena, that of genetics. Around 1990 it was discovered that about half of the mem-

< 165 >

bers of three generations of an extended family in England, known as the KE family, were affected by a severe disorder of speech. It seemed to be of genetic origin, because it was apparent from the very first efforts of the affected child to speak and persisted into adulthood.[67] It was later shown to be due to a point mutation on the FOXP2 gene on chromosome 7.[68] It has nothing to do with foxes; *FOXP2* stands for "forkhead box P2." In order to learn to speak normally, you need two functional copies of this gene. The discovery of this gene created something of a furor in linguistics. Some hailed it as the language gene,[69] which gave Prometheus the gift of universal grammar. This idea is no longer widely held, but FOXP2 may have had something to do with the emergence of the ability to speak.

Perhaps it was the gene that somehow introduced vocalization into the mirror system, so that speech could develop as an intentional, learnable system. Brain imaging has shown that Broca's area, which is critical to the production of speech, remains largely inactive in members of the KE family affected by the FOXP2 mutation when they generate verbs but is activated normally in their unaffected siblings.[70] What's more, the activation in the unaffected members was on the left side, while the sparse activation in the affected members was scattered on both sides of the brain, seemingly favoring neither. Under normal conditions, then, the human form of the FOXP2 gene appears to have somehow admitted control of vocalization into Broca's area only on the left side.

The gene itself, though, or variants of it, is not restricted to humans, and it goes back a long way in mammalian evolution. Indeed, it differs very little from that in the mouse. Nevertheless, two changes in a region of the gene, exon 7, have occurred since the split between humans and chimpanzees, suggesting that one or both of these changes may have been somehow responsible for enabling humans, but not chimps, to talk. Estimates based on mathematical assumptions suggest that the most recent mutation occurred within the past one hundred thousand years, or even as late as around forty thousand years ago,[71] suggesting that it was not present in Neander-

< 166 >

tals. So if the recent mutation was really critical to the emergence of speech, the Neandertals may have remained speechless. Indeed the implied dates once again conform to the Chomskyan view that human language emerged in a big bang less than a hundred thousand years ago—although please remember again that speech is not the same as language.

But the plot thickens. Contrary to the mathematically based estimates, the human mutation is also present in the DNA of a forty-five-year-old Neandertal fossil, implying that it goes back at least four hundred thousand years to our common ancestry with the Neandertals.[72] That estimate may be threatened in turn by the evidence, mentioned in chapter 2, evidence that early humans mated with Neandertals. Perhaps a talking *sapiens*—Prometheus even—proved irresistible to some gesturing Neandertal of the opposite sex, and a cross-breed was born carrying the mutation, which was eventually seduced into the rest of the population. But this would not have applied to those early humans who remained in Africa, because they had no contact with Neandertals. Another possibility is that the mutation in question has nothing to do with speech and that a mutation outside of exon 7 occurred within the past two hundred fifty thousand years and is unique to humans.[73] On the other hand, all of this may be wishful thinking, born of a desire to distance ourselves from our Neandertal cousins.

In any case, FOXP2 does seem to play a role in vocalization in species other than humans. Songbirds also have the gene and are less able to imitate song if it is knocked out,[74] and replacing the FOXP2 gene in the mouse with the mutated version found in the KE family makes the poor animal less adept at motor learning. As though to compensate, inserting the normal human variant into the mouse makes a qualitative difference to its ultrasonic squeaks, although it did not transform the squeaking into speaking. The authors of this study were nevertheless bold enough to suggest that their results could bear on the evolution of speech, if not of language itself. Extrapolating from mice to humans, they write:

< 167 >

However, since patients that carry one nonfunctional FOXP2 allele show impairments in the timing and sequencing of orofacial movements, one possibility is that the amino acid substitutions in FOXP2 contributed to an increase in fine-tuning of motor control necessary for articulation, i.e., the unique human capacity to learn and coordinate the muscle movements in lungs, larynx, tongue and lips that are necessary for speech. We are confident that concerted studies of mice, humans and other primates will eventually clarify if this is the case.[75]

To complicate matters further, though, it now transpires that FOXP2 is much more than a speech gene and probably influenced the evolution of speech only indirectly. It is widely involved in development, not only in the brain but also in the gut and lungs.[76] It is what is known as a *transcription factor*, which means that it regulates the operation of other genes, so it may be these other genes that play the critical role in speech. The current view seems to be that the system governed by FOXP2 has to do with plasticity and motor learning in a wide range of animals. Learning to speak may be just one manifestation of this broader portfolio. To unravel this system will obviously require a lot more work, but at the present stage of inquiry it seems clear that the advent of speech, or language for that matter, is unlikely to have been the outcome of a single mutation.

The saga of FOXP2, especially in relation to Neandertals, remains of intense interest, in part because the Neandertals had brains at least as large as ours and had at least some of the trappings of intelligent behavior, including the manufacture of tools. As noted above, there is perhaps a desire to see them as inferior to our own species, as though to assuage the sense that we might have been responsible for their demise. Humans have a sorry record of genocide. To suppose that the Neandertals brought about their own demise through an inability to speak, or more dramatically through an inability to think as we do, might bring a small element of solace. But on the other hand there is a fellow feeling for the Neandertals, so much like ourselves that it would be nice to know we could have chatted with them.

< 168 >

Whatever the outcome of the FOXP2 story, it seems increasingly unlikely that it will support the idea of a big bang that transformed language and cognition uniquely in our species. It may turn out that FOXP2 was another straw to be clutched in the desire to prove the uniqueness and superiority of our species—and probably not the last straw.

Were Neandertals Human after All?

In chapter 2 I raised the question whether the Neandertals and their fellow travelers the Denisovans were indeed species distinct from *Homo sapiens*; I noted that some modern-day humans carry a small fraction of Neandertal and Denisovan genes. The Neandertals in particular, though, have generally been regarded as grotesque subhuman brutes. John E. Pfeiffer once remarked that when Neandertal skulls were first discovered in 1856, they "came into the world of the Victorians like a naked savage into a ladies' sewing circle."[77] But a few of the ladies of the past seem to have mated with them, which suggests that they were not a different species. Paola Villa and Wil Roebroeks refer to the "human superiority complex" in attitudes to the Neandertals and suggest that "complex processes of interbreeding and assimilation may have been responsible for the disappearance of the specific Neandertal morphology from the fossil record."[78] So perhaps we've been with them, and they with us, all along.

The Swedish physicist and polymath Sverker Johansson argues, contrary to Chomsky and other "big bang" theorists, that the Neandertals must have had some form of language, possibly gestural, although he also suggests that they were probably capable of speech at some level.[79] Dan Dediu and Stephen Levinson are even more dismissive of the "saltationist" view of language, suggesting that speech and modern language probably go back at least half a million years to the common ancestor of modern humans, Neandertals, and Denisovans.[80] A rock engraving recently found in a cave in Gibraltar was evidently carved by Neandertals and has been taken as evidence

< 169 >

for "abstract thought and expression through the use of abstract forms."[81] (The engraving suggests they may have played tic-tac-toe.) We should stop saying unkind things about the Neandertals and welcome them to the sewing circle.[82]

Whatever the outcome of the FOXP2 story, or of attempts to reconstruct the Neandertal vocal tract, the best guess is that speech was not the outcome of any single mutation. It is nevertheless possible that our forebears who migrated from Africa some sixty thousand years ago and eventually populated the globe were somewhat distinctive in possessing a greater power of speech than their earlier cousins. One possibility is that speech may have been a cultural invention that emerged gradually, rather than the result of a sudden genetic change. Language had probably developed both gestural and vocal components in large-brained hominins, including Neandertals and Denisovans, but it may have been the early humans who hit upon the idea of making vocalization the dominant mode. Even as an invention, though, speech may well have favored the selection of genetic and anatomical changes that sharpened the flexible control of voicing. Other examples of inventions with profound impacts on social life have led to the selection of genetic mutations that enhance them. It sometimes seems to me that kids these days are already born to understand the electronic age. And watching teenagers texting on their cell phones makes me wonder whether we are returning to gesture as the dominant mode. In terms of bodily expression, language may be reduced to a twiddling of thumbs, with electronics doing the rest.

It may have been a shift to autonomous speech, rather than the emergence of language itself, that was responsible for the "great leap forward" of the past hundred thousand years, discussed in chapter 2— if indeed such a leap truly occurred. As suggested earlier, speech freed the hands, making way for the advances in technology that have so transformed the planet. Again, speech also enhances pedagogy, enabling people to explain manual techniques of toolmaking and tool use while at the same time demonstrating them. And speech is also an end product of conventionalization, leading to symbols that are

< 170 >

energy efficient and largely arbitrary. The arbitrariness of words in fact serves social functions, because it binds individuals together in communities while at the same time keeping different communities apart. To outsiders, a foreign language is like a barrier. Speech mutates at an astonishing rate, with the six thousand or so languages in the world, driven in part by a natural human tendency to tribalism. Perhaps the world would be safer if we all took up sign languages instead. Sign languages also become conventionalized but retain an element of pantomime that makes communication across cultural boundaries more transparent—although perhaps only marginally so.

The next question, though, is how language, whether spoken or signed, was shaped to enable the outward expression of episodes and, more expansively, stories. This is essentially the problem of grammar. But before we plunge into that, it's a nice sunny day and I'm off to the beach.

< 171 >

9

HOW LANGUAGE IS STRUCTURED

My voice goes after what my eyes cannot reach,
With the twirl of my tongue I encompass worlds, and volumes
of worlds.

Walt Whitman, *Song of Myself*

Let's begin with the basic requirements. Our internal thoughts can be considered to be made up of concepts, which are in turn derived from the way in which we parse the world. We see objects such as trees and helicopters and ice cream cones, observe or carry out actions such as hopping and screaming and flirting, feel emotions such as fear and nostalgia and jealousy. To convey our thoughts, then, we need outward bodily-produced symbols, or words, to refer to these concepts so that we can transmit them to other people even if the concepts themselves are not represented in the immediate environment. That is, we can tell stories about real and imaginary worlds. These symbols we use may be visible signs, as in sign language, or they may be sounds, as in speech, or they may be specific shapes, as in writing.

Such a system allows for displacement, the ability to refer to the nonpresent. If I'm to tell you about something that happened yesterday, or in 1945 (yes, I can remember that far back), I need symbols that evoke the objects, actions, people, and so forth that populated those past events. This in turn requires the assumption that we share the same concepts, so that when I say *ostrich* or *gluon* or *char-à-banc*, I invoke in your mind the entities that those words represent. The same applies to accounts of future events or purely imaginary ones. Con-

< 172 >

cepts are themselves nonverbal and are not restricted to humans, although a hypothesis generally attributed to the linguist Benjamin Lee Whorf[1] goes so far as to claim that our concepts are determined by the words that refer to them, implying that nonhuman animals would have at best a paucity of concepts. The strong form of the Whorfian hypothesis has been largely discredited, although some concepts, such as individual colors, may well be refined or circumscribed by the words we use to label them with.

In the industrialized world, at least, our densely packed brains house a vast number of concepts. A conservative estimate of the number is the number of words we have to refer to them, and as we have seen our vocabularies extend to tens of thousands. Steven Pinker gives an estimate of fifty thousand, based on a college dictionary, and the linguist James Hurford thinks his own vocabulary is about seventy thousand. And there are of course words that don't appear in some dictionaries, such as the names of individual people, countries, towns, and pet rabbits. These personal words refer to concepts that inform our individual lives.

As suggested in chapter 7, the symbols we use to evoke concepts in others may have originated in gesture, the configurations of our hands and bodies to represent objects and actions. People have also long used other representational symbols, such as drawings or figurines, to evoke objects, animals, or people, and with modern technology we can use photographs and videos to capture and store past events and the entities that populated them. In evolutionary terms, though, the symbols we use in ordinary language probably originated in mimes, bodily depictions of shapes and actions. Through time, mimes would have been gradually adapted or replaced with more efficient and streamlined forms, at the expense of physical resemblance or iconicity. As suggested in the previous chapter, voiced sounds provided the dominant way to achieve this. The linking of symbols, whether manual or vocal, with concepts is established and maintained through convention rather than physical resemblance—the process called conventionalization. The conventionalized symbols that we use to refer to objects, actions, or qualities, whether in

< 173 >

speaking or signing, are what we call *words*,[2] although we also use other kinds of words, such as articles, prepositions, and modifiers. These arise in the context of sentences.

The use of symbols does not apply only to language. Perhaps it was the freeing of the hands that led to the making of physical artifacts as cultural emblems. A period of life on the beach or lakeshore, during the so-called aquatic phase discussed earlier, may have contributed. Marcos Duarte notes that from around two hundred thousand years ago or more, our forebears used shells and ochre for decoration and symbolic representation, perhaps reflecting a longtime tradition of extracting food from around shorelines.[3] Shells and ochre used as bodily ornamentation is widely documented in ancient cultures from Mesopotamia, Egypt, and Greece to ancient Rome—and indeed to my compatriots the New Zealand Maori. Noblewomen in ancient Rome were especially fond of using ochre in their makeup, which inspired the Roman playwright Plautus (254–184 BCE) to remark that "a woman without paint is like food without salt."[4] The early European explorers called Native Americans "redskins," not because of the color of their natural skins but because of the widespread use of ochre. The influence of ochre extends no doubt to the widespread use of lipstick—although I believe that is made from other substances. Duarte observes that shells and ochre are also associated with the Neandertals, again suggesting that we may not be quite as distinct from them as we like to think we are.

Using conventionalized symbols that are for the most part non-iconic, such as spoken or printed words, requires the facility to learn to associate them with the concepts that we know, since we can't rely on resemblance. Although words may have begun as mimed representations, or in some cases as sounds that mimic reality, the child raised in a speaking environment is normally presented with words whose shapes or sounds have no obvious relation to what they stand for. Children are nevertheless extraordinarily proficient at making the connections. From the age of about two, English-speaking children are said to learn new words, and what they stand for, at the rate

< 174 >

of about ten per day, until they reach a vocabulary of around sixty thousand.[5]

Learning to attach arbitrary symbols to concepts, though, is not unique to humans. The bonobo Kanzi uses a keyboard with over three hundred symbols representing objects and actions, and points to them to make simple requests. The symbols are deliberately constructed to bear no physical resemblance to what they stand for. We also saw in chapter 2 that Kanzi has a moderate ability to understand spoken speech, and the border collie Chaser is said to have a receptive vocabulary of over a thousand spoken words. For these dogs, the words give no physical cue as to what they stand for; their referents must be learned. Parrots, dolphins, and sea lions can also build vocabularies of about the same order.[6] The border collies Rico and Chaser quickly learn new words through exclusion; given a new word and a selection of objects, they pick out an object they haven't seen before and attach the new word to it. The felicitous capacity to attach arbitrary labels to objects is called *fast mapping*.[7] Children are especially good at it, but smart animals like border collies may not be far behind. The main difference is that the collies can't actually produce the words.[8]

Arbitrariness may actually be the rule rather than the exception in animal communication, at least in the vocal domain. The classic work of Robert M. Seyfarth and Dorothy L. Cheney showed that vervet monkeys made distinct alarm calls to signal the presence of different predators (leopard, eagle, and snake), but the calls themselves bore no resemblance to the sounds made by those predators.[9] The same turns out to be true of many different species of monkey,[10] and birdsongs generally give no direct indication of what they are about or refer to—with the exception of birds that mimic, such as the parrot and mynah bird. For the most part, though, the calls by these various animals seem not to be intentional, but they do make the point that signaling in nonhumans is for the most part noniconic in form.

To be sure, there are some two orders of magnitude difference between the sheer size of vocabularies that we humans can build and

< 175 >

the sizes of those acquired by smarter nonhumans, but we can again invoke our Darwinian mantra: "The difference in mind between man and the higher animals, great as it is, certainly is one of degree and not of kind." The brain is an association machine that readily links otherwise unrelated inputs, and the advanced capacity of the human brain to do this may be at least partly a matter of its size. As I noted in chapter 1, the human brain tripled in size relative to that of the chimpanzee, and much of this increase took place during the Pleistocene, when the genus *Homo* emerged. This increase may have been an adaptation to the demands of the social mind, which no doubt included the vast memory capacity required by the store of concepts and words that must be attached to them, as mimetic culture gradually gave way to more conventionalized representations. Perhaps there was also an increase in association strength. Words become so firmly attached to concepts that one can scarcely think of a concept without eliciting the word, and vice versa. Even so, they can become detached, as when we forget the name of someone we know quite well and can easily picture in the mind's eye. It also happens, but more rarely, that we know a name but can't quite picture to whom it belongs.

Of course concepts themselves can be abstract as well as concrete. We speak of such things as love, prudence, fairness, obnoxiousness, creativity—and even, paradoxically, "the love that dare not speak its name."[11] We use words to label these abstract concepts in much the same way as we use concrete words to label the physical entities of the world, and indeed we often use concrete metaphors to express more abstract ideas:

> Happiness is a warm puppy.
> Life is a bitch.[12]
> Time like an ever-rolling stream.
> Love is a many-splendored thing.

And although we use words to refer to concepts, we also understand words themselves as concepts. We can regard them as specific manual signs, or as patterns of sound, or printed shapes, and we can

< 176 >

classify them as nouns, verbs, adjectives, and the like. We have words for words themselves; it's a recursive jungle in there.

Episodes and Sentences

But there is much more to our mental experiences than just concepts. We perceive, remember, invent, and imagine *episodes*, slices of experience during which something happens. Our ability to do so is impressive. In experiments run in our laboratory here, participants are asked to recall 110 past events of personal significance and identify from each of them a person, an object, and the location at which the event occurred. The people, objects, and locations so remembered are rearranged by the experimenter into new triplets not previously involved in the same event, and the participants asked to imagine a future event based on each triplet. They are then placed in a brain scanner to find out which parts of the brain are activated while they undertake these acts of memory and imagination. (Needless to say, it's the hippocampus that features prominently in both directions of mental time travel.)[13] Our undergraduates, who bear the burden of this research, seem to have little difficulty, and the experimental tasks really only scratch the surface of the human ability to remember the past or imagine new scenarios.

Of course we don't remember everything, and the émigré Czech writer Milan Kundera reminds us that our waking hours are witness to far more events than we can possibly remember.

> The fundamental given is the ratio between the amount of time in the lived life and the amount of time from that life that is stored in memory. No one has ever tried to calculate this ratio, and in fact there exists no technique for doing so; yet without much risk of error I could assume that the memory retains no more than a millionth, a hundred-millionth, in short an utterly infinitesimal bit of the lived life. That fact too is part of the essence of man. If someone could retain in his memory everything that he had experienced, if he could

< 177 >

at any time call up any fragment of his past, he would be nothing like human beings: neither his loves nor his friendships nor his capacity to forgive or avenge would resemble ours.[14]

He does exaggerate, I think, since a hundred-millionth would amount to only a few seconds' worth. Even so, it's true that much of our conscious lives slips by without lasting trace. We do have a tendency to think that we remember more of our life than we actually do, but that's simply because we have forgotten much of it!

Nevertheless, we do seem to house an impressive supply of remembered episodes, along with the inexhaustible capacity to juggle our memories into novel happenings, projected or simply imagined—or for that matter falsely remembered. Most of us are capable of constructing our autobiographies, which in some cases run to several volumes. A striking example is the New Zealand author Janet Frame's three-volume work recording her life in extraordinary detail, despite her having endured many bouts of electroconvulsive therapy as a psychiatric patient.[15] She also wrote a number of highly acclaimed novels. As I suggested in chapter 4, it may be the unbounded nature of episodes, remembered or imagined, that explains why language itself is often taken to be unbounded. We need an unbounded system to be able to transmit information about an unbounded world.

The episodes in our lives, whether actual or imagined, involve combinations of concepts, and these combinations stretch the storage requirements much further. It is through sentences that we describe episodes and so share the experience of those episodes with others. As every schoolchild knows—or used to know—every sentence must have a verb, to represent that action. Language is sublimely adapted to arrange the words representing the concepts into sentences describing episodes—what Pinker called "who did what to whom, when, where, and why"[16]

Grammar, then, may derive essentially from the problem of relaying experienced or imagined episodes through combinations of words. The most basic elements of episodes, the events of our lives, are objects and actions, which can generally be parsed into what have

< 178 >

been termed *agent*, *action*, and *patient*, or who did what to whom—as in "Natasha teased Lena."[17] In language these are represented by subject (S), verb (V), and object (O), respectively, requiring just two kinds of words, or parts of speech—nouns and verbs. In a simple mime, these can be played out much as they occurred in an actual or imagined event. In miming a cow jumping over the moon, for example, one might depict the cow with a specific shape of the left hand, say, the moon as a shape of the right hand, and the act of jumping by a movement of the left hand over the right. (Go on, do it, but don't knock anything over.)

Some events, like that hyperactive cow jumping over the moon, can be mimed holistically, so there is little sense of an ordering among the elements themselves. In the course of conventionalization, though, it is expedient to separate the elements and mime them sequentially. In the example of the cow jumping over the moon, depicting the cow independently of its spectacular leap allows that symbol to be used in other contexts, such as a cow being chased by a dog or simply contentedly eating grass. Similarly, a depiction of the jumping movement can be separated from the cow and also used in other contexts, such as that of a high school sports day or a jumping flea.

An example of this kind of segmentation is provided by the sign language invented and elaborated by deaf children when schools for the deaf were created in Nicaragua in 1979. Over time, Nicaraguan Sign Language changed from a system of holistic signs to a more combinatorial format. For example, one generation of children was told a story of a cat that swallowed a bowling ball and then rolled down a steep street in a "waving, wobbling manner." The children were then asked to sign the motion. Some indicated the motion holistically, moving the hand downward in a waving motion. Others, however, segmented the motion into two signs, one representing downward motion and the other representing the waving motion, and this version gained currency after the first cohort of children had moved through the school.[18] Thus the move toward segmentation is a natural process in the development of communication systems, allowing for greater efficiency so that the segmented signals can be used in dif-

< 179 >

ferent contexts, and perhaps also enhancing the fidelity of transmission, just as digital systems in audio and video recording have replaced analog ones. Computer simulations have shown how cultural transmission can change a language that begins with holistic units into one in which sequences of forms are combined to produce meanings that were earlier expressed holistically.[19]

Once a system is compartmentalized into discrete units, the ordering of those units can become critical and depends largely on convention, as communities settle on common rules. To use a well-chewed example, the sentence "Man bites dog" means something very different from "Dog bites man"—the one is news, the other simply a personal misfortune. In spoken English the default word order is subject-verb-object (SVO), but this is not the most common order. Among the world's spoken languages, the predominant order is SOV, with the verb at the end, and it may well be that this is the most "natural" sequence in miming simple events—first introduce the cast, then display the action. Murray Gell-Mann and Merritt Ruhlen have argued that the ancestral human language, which they propose arose in Africa more than fifty thousand years ago, was an SOV language.[20] In English, as noted, words are placed in the order subject-verb-object (SVO). Nevertheless, all possible combinations exist among the world's spoken languages today, although among spoken languages there seem to be only four that are OSV languages. In case you are traveling this year, they are Warao in Venezuela, Nadëb in Brazil, Wik Ngathana in northeastern Australia, and Tobati in West Papua New Guinea.

Word order in English (and many other languages), though, is flexible and can even be reversed, as when we use the passive, such as "the dog was bitten by the man" or "the moon was jumped over by the cow." And then there's the pernicious influence of *Time* magazine, where, according to Wolcott Gibbs, "backward ran sentences until reeled the mind."[21] Word order may be less critical in sign languages because they retain a holistic element; in British Sign Language, however, the default order is OSV, so that it joins the rare spoken OSV languages mentioned above.

< 180 >

Some languages don't rely on word order to distinguish subject, object, and verb. Nouns are inflected according to their status (subject, object, indirect object) and verbs according to their tense (and other things), so the meaning of a sentence can be discerned more or less regardless of how the words are ordered. I remember from my schooldays having to learn the various declensions of Latin nouns and conjugations of Latin verbs that indicate their roles in sentences. (I have no idea why nouns are declined and verbs conjugated.) I gave further examples of the varied ways in which sentences are formed in chapter 1. Languages in which word order plays little or no role are called *scrambling languages*, an extreme example being the Australian aboriginal language Walpiri. You can shuffle the words without affecting the meaning.

Grammaticalization

In the course of development, languages have advanced well beyond combinations of nouns and verbs and include many words that do not refer to actual content. These are called function words and include articles, such as *a* and *the*, prepositions such as *at*, *on*, and *about*, and auxiliaries such as *will* in "they will come." Function words almost certainly have their origins in content words, and the process by which content words are stripped of meaning to serve purely grammatical functions is known as *grammaticalization*.[22] A classic example is the word *have*, which progressed from a verb meaning to "seize" or "grasp" (Latin *capere*) to one expressing possession (as in "I have a pet rabbit"; Latin *habere*), to a marker of the perfect tense ("I have gone") and a marker of obligation ("I have to go"). Similarly, the word *will* progressed from a verb (as in "Do what you will") to a marker of the future tense ("they will laugh"). Similar examples can be found in other languages, as in the Michel Sardou song "Je vais t'aimer" ("I will love you").

Another example comes from the word *go*. It still carries the meaning of travel, or moving away from a location, but in sentences like

< 181 >

"we're going to have a party" it has been bleached of content and simply indicates the future. The phrase *going to* has been compressed into the form *gonna*, as in "we're gonna have a party" or even "we're gonna party." We even combine the different meanings, as in "I'm gonna go." The battle cry of the Chicago White Sox, I'm told, is "Let's go go-go."

In some languages, function words were compressed into content words to create inflections and can even seesaw from one to the other. For example, in Latin the original inflection that marked the future tense, as in *cantabo* ("I will sing") was dropped because it was too close to the past imperfect, as in *cantabam* ("I was singing"), and replaced by use of a function word, as in *cantare habeo* ("I have to sing"). But in subsequent developments the function word was re-inserted as a full inflection, as in the Italian *cantero* or French *je chanterai*. Grammar is thus at the mercy of common usage, creating conventions that are carried by culture.

Different parts of speech are also formed by altering existing elements, starting with nouns and verbs. For example, nouns can be created from verbs by adding suffixes (*improve* → *improvement*; *elect* → *election*), adjectives from nouns (*disaster* → *disastrous*; *grief* → *grievous*), adverbs from adjectives (*grievous* → *grievously*, with the *-ly* suffix a compression of *-like*). In the interests of greater economy, grammaticalization also leads to the embedding and concatenization of phrases. For example, the statements "te pushed the door" and "the door opened" can be concatenated into "he pushed the door open." Statements like "my niece is fickle with money" and "my niece paid far too much for her wardrobe" can be concatenated by embedding the first in the second: "My niece, who is fickle with money, paid far too much for her wardrobe." One can also alter the priority of the two statements by reversing the embedding: "My niece, who paid far too much for her wardrobe, is fickle with money."

The evolution of language is a process whereby communication is enriched while at the same time achieving greater precision and economy. The earliest forms of language may have resembled pidgins, which in more modern times were invented as makeshift languages

< 182 >

enabling European traders and colonizers to communicate with indigenous peoples. They have little or no grammar and rely heavily on nouns and verbs, resulting in phrases that are repetitive and inefficient. In Solomon Island pidgin, Prince Charles is known as *pikinin belong Missus Kwin* and his late wife Princess Diana as *Meri belong pikinini belong Missus Kwin*. When they divorced, Diana was known as *this fella Meri he Meri belong pikinini belong Missus Kwin him go finish*.[23] In English she was simply *Princess Di*—although, to be fair, her belongingness was tacitly known to all and could remain unspoken.

Paul Hopper and Elizabeth Traugott suggest that grammaticalization is unidirectional, creating changes that are seldom if ever reversed. Since this depends on culture, it can explain why languages have diverged not only in vocabulary but also in structure. Languages have diverged into mutual incomprehensibility, partly through different cultural imperatives, partly through the sheer randomness of cultural change, and perhaps partly to create impermeability. Language is a fortress.

Our Spatial Heritage

Although different languages have come to be structured very differently, the generative structure of language, what Chomsky called "discrete infinity," may derive from the unbounded nature of our spatial experiences, as suggested earlier. This derives in turn from the boundless nature of the spatiotemporal world—for all practical purposes, space and time extend infinitely. It is this boundlessness, which itself is nonlinguistic and indeed not restricted to humans, that underlies the generativity of language.

The idea that the structure of language derives from our spatial heritage is not new and has been developed by a number of linguists, notably Paul Deane.[24, 25] His argument is partly based on neurology, particularly on the role of a brain area called the inferior parietal lobule (IPL). In the primate brain this has mostly to do with the perception of space. In humans, though, the left IPL is involved in lan-

< 183 >

guage, while the right IPL retains a purely spatial agenda—in most people at least. Deane notes evidence from brain imaging that the left IPL is activated when people imagine grasping a three-dimensional object with the right hand. These observations suggest an evolutionary scenario in which both halves of the IPL were initially concerned with spatial mapping but the left half was later adapted for grasping, and then for spoken language—a scenario also consistent with my theme, developed in chapter 7, that language evolved from manual grasping.

The importance of space in language takes us back to the hippocampus, the brain's central station for spatial thinking and memory. As this seahorse-shaped structure nestled in your brain will again remind you, it allows us not only to understand the spatial world surrounding us but also to reconstruct past episodes in time and space as well as imagine future ones. Time and space are interwoven in our very speech; the neuroscientist (and recent Nobelist) John O'Keefe notes that most prepositions in English, such as *at, about, across, against, among, along, around, between, from, in, through*, and *to*, apply to both time and space and apply even to logical expressions that are symbolic rather than spatial, as in "A follows from B" or "the argument against A is B."[26] Anticipating Deane, O'Keefe also suggests that the hippocampus and adjacent structures gained an asymmetry; he notes that damage to the right mesial temporal lobe (which includes the hippocampus) results in amnesia for episodes coded in visuospatial terms, while left-sided damage results in amnesia for episodes coded verbally.[27] In a 2012 review of the effects of hippocampal damage, Melissa Duff and Sarah Brown-Schmidt conclude that the hippocampus "is a key contributor to language use and processing."[28]

In outlining his spatial theory of the origins of grammar, Paul Deane comments that "if grammatical structure is a metaphorical projection of this system, then linguistic expressions are being processed as if they were physical objects."[29] Some have suggested that the manufacture and use of tools may also have contributed to the structure of language, and indeed manufactured objects can have all of the recursive complexity of language itself. The influence of tools

< 184 >

may have arisen during the Pleistocene. Dietrich Stout and Thierry Chaminade describe experiments in which brain activity was recorded while people made tools and while they watched tools being made.[30] They focused on toolmaking techniques from two periods of toolmaking in hominin evolution: the Oldowan, dating from around 2.6 million years ago and representing the earliest period of stone tool manufacture, and the late Acheulian, appearing round seven hundred thousand years ago. Acheulian toolmaking saw the emergence of extra activity in the frontal lobes in regions associated with language. Acheulian toolmaking in particular begins to include "grammarlike" attributes. One such attribute is the combining of parts, as in hand axes requiring the hafting of handles onto a blade. Hafted hunting tools go back some five hundred thousnd years, probably to the common ancestor of *Homo sapiens* and the Neandertals.[31] Perhaps hafting marks the origin of Chomsky's Merge operation, but placing it in practical life rather than in language. Ian Tattersall writes that "the invention of the hand-axe must have represented—or at least reflected—a cognitive leap of some kind,"[32] although not so substantial a leap, in his opinion, as the one that finally gave us language itself. The use and manufacture of tools would also have added to the mimetic repertoire and the flexibility of hand and arm movements.

Another grammarlike feature of toolmaking is its hierarchical nature, with subgoals that must be met within overarching goals. For example, in the removal of stone flakes, various "subroutines" such as edging, thinning, and shaping need to be carried out before a flake is removed. One consequence of this is a separation in time between related operations—a subroutine must be completed before the toolmaker can return to the original goal. Likewise in language, one must retain information from the beginning of a sentence while a subordinate clause is processed before the rest of the sentence completes the meaning. That last sentence illustrates the very point: the subordinate clause "while a subordinate clause is processed" must be processed before the sentence grinds to its conclusion. Got it? This analysis provides yet another extension of the idea that language evolved from manual gesture and that even grammar may owe something of its

< 185 >

origin to tool construction.[33] Stout and Chaminade also suggest that Acheulian toolmaking demonstrates an intentional aspect not present in the earlier Oldowan phase. They write: "Evidence of intention attribution during the observation of stone tool-making provides support for a 'technological pedagogy' hypothesis, which proposes that intentional pedagogical demonstration could have provided an adequate scaffold for the evolution of intentional vocal communication."[34]

Perhaps, though, the influence was the other way round, with language influencing the way mechanical tools were developed. Grammar may well owe its origin to social rather than mechanical forces, as our hominin forebears settled into the "cognitive niche" with its social complexity and increasing need to plan, record, and share. Perhaps, too, the relatively late appearance of multipart tools in the Pleistocene was a consequence of language itself shifting from a manual mode to a vocal one, freeing the hands for more elaborate toolmaking, although it may have been even later, after the arrival of *Homo sapiens* some two hundred thousand years ago, that the transition was complete.

But in a broader perspective, language itself can be regarded as a tool. As I mentioned in chapter 3, recent authors such as Daniel Dor[35] and Daniel Everett[36] consider that language to be a form of technology rather than a mode of thought. Manufacture, language, and other pastimes such as music and art can all be considered part of the human response to a postarboreal habitat. Adaptation is a two-way process, involving changes both to the organism and to the environment it inhabits. In the case of our own species, the balance seems to lie in the latter, the changes we have wrought on the environment. We and our forebears have worked, ultimately on a massive scale, to adapt the environment itself to our own ends. The imprint of manufacture is evident in modern cities and increasingly sophisticated modes of transport, not to mention communication itself. We are awash in words, not only through the everyday chatter of gossip and commercial radio but also through the vast and ubiquitous changes in the mode of communication itself. Manufacture and language are

< 186 >

at one in the tweets of teenagers on their smartphones, wagging their thumbs in preference to their tongues.

Abstraction

In this chapter I have argued that language emerged in the context of episodes, whether remembered, planned, or imagined. The sentence corresponds at least roughly to the episode. Episodes also form the basis of stories, which in their simplest form are sequences of episodes. Again at the most basic level, nouns and verbs correspond to the objects and actions of our episodic lives. We tell stories about the interrelations between people, things, and happenings.

Modern linguists will see this as overly simplistic, but my aim is to provide an account of how language evolved and not of how it is understood in contemporary linguistics. To linguists, parts of speech are understood in terms of their roles in sentences and not in terms of the sorts of things they refer to. This is in itself a product of abstraction, whereby words themselves are taken a further step away from what they represent and become playthings of the mind. Once both iconic representation and meaning are removed, they can be paradoxically at once extended and sharpened. I remember in my schooldays being taught matrix algebra, a branch of mathematics that is elegant and precise, but I gained no sense of how it might apply to anything in the real world. Later, as a student of psychology, I discovered a statistical technique called factor analysis and saw that it was built on matrix algebra. This came as something of a revelation, and for a while I worked on the development of factor models as a way of understanding aspects of the human mind.

In much of linguistics, language itself has become an exercise in symbol manipulation, a development largely inspired by Noam Chomsky. The abstraction of language away from reference to the natural world no doubt underlay Chomsky's view that language is internal to the mind and could not have evolved through natural selection. Chomsky's "unbounded Merge" is an example of an ab-

< 187 >

stract principle that may lead to more advanced understanding of how language works. This is not to say, though, that abstract thinking arose independently of natural selection or as a catastrophic leap at some point in human evolution. It is more likely a product of gradual conventionalization, along with an ingenuity that can lead to the formulation of abstract truths with potentially broad application. Even unbounded Merge may have application well beyond language itself. The computer scientist Alistair Knott derives the essence of Chomsky's Minimalist Program from simple sensorimotor actions such as grasping a cup! In the final paragraph of his book he writes: "The universal, biologically given capacity for learning languages postulated by Minimalism exists, and its operation can be described using the kind of theoretical machinery which is posited by Minimalism. However, this capacity is not a language-specific one; Minimalism's syntactic analyses are in fact analyses of general properties of the human sensorimotor system, and of memory representations which reflect these properties."[37] This is an endorsement of the gestural origins of language, and indeed of the increasing view that language is fundamentally an embodied system rather than one based on the manipulation of abstract symbols.

Nevertheless, the emergence of abstract symbols through conventionalization and the streamlining of language may well have had an influence on the way we think. At the beginning of chapter 6 I noted William James's suggestion that there are two kinds of thinking—narrative thinking and reasoning. Reasoning may well have emerged along with the increasing use of abstract symbols separated from resemblance to what they represent. Extreme examples include algebra and computing languages, along with the highly mathematized expressions sometimes used in modern linguistic analysis. But most people still prefer stories to logic. And it was stories, I think, that came before reason.

< 188 >

10

OVER THE RUBICON

And so, when he was come to the river Rubicon, which was the boundary of the province allotted to him, he stood in silence and delayed to cross, reasoning with himself, of course, upon the magnitude of his adventure. Then, like one who casts himself from a precipice into a yawning abyss, he closed the eyes of reason and put a veil between them and his peril, and calling out in Greek to the bystanders these words only, "Let the die be cast," he set his army across.[1]

Plutarch, *Life of Pompeius*

It's been quite a long journey, and that river has been difficult to cross, but I think we can now begin to piece together the sequence of events that led to language. We need first, though, to clarify again what we mean by *language*. The approach taken in this book departs from that of Chomsky and others in that language is here regarded as a communication system and not as a system of thought—E-language rather than I-language. Chomsky himself regarded external language as relatively trivial and uninteresting, with internal language as the real challenge. Here this is reversed, which "turns the Chomsky proposal on its head," as Daniel Dor put it. And, pace Chomsky, external language is far from a walk in the park. It is, after all, a human phenomenon, allowing us to communicate freely with each other about our ideas and experiences, but not with other animals—nor other animals with each other, except in very limited ways.

If language is the sharing of thoughts and experiences, this does

< 189 >

not mean that we can ignore thought itself. Language must be shaped to map onto our thoughts and make them accessible to others. The important properties of thought go back in time well before the emergence of our species and may indeed have ancient roots. Foremost among these is experience of events in time and space. Our knowledge of space has ancient roots. Brain regions, critically involving the hippocampus, have evolved to record locations in space and even permit the replaying and forward planning of movements in spatial environments. To a limited degree, at least, this may be as true of rats as of humans. It is in space that we wander, grasp, frolic. Our minds, adapted to space, do likewise. Our understanding and manipulation of space may have even set the stage for grammar, allowing us to manipulate words as a general orders his troops, or a chess player her pieces on the chessboard.

To the extent that language has universal properties, it probably owes them more to the common nature of experience than to language itself. We all live in similar spatiotemporal worlds, inhabited by things, people, and various artifacts of our own making, and we live lives of similar length. We are size-scaled in like manner, and differently from ants or whales. Given normal development, we all view the world at the same angle and move at roughly the same speed, although fast cars and airplanes may make a difference. Of course there are quite wide environmental differences, such those between African Bushmen and New York traders, but overall we are scaled to similar environments, and our notions of what the world is are probably roughly similar across the globe, notwithstanding extremes of climate and geography. Moreover, our minds were largely shaped at a time when we shared environments that were much more similar than they are now.

What Chomsky called *universal grammar* may therefore depend more on how long we and our forebears have inhabited the world and reacted to it than on some new internal program called unbounded Merge. As I explained in chapter 4, even rats may generate different spatial maps in a flexible and recursive fashion—or so their hippocampi tell us. Generativity, then, derives not so much from the

< 190 >

method of sharing our experiences as from the experiences themselves.

The manner of internally representing the contents of space-time—the objects and actions that fill lives—probably also goes far back in evolution. Remember those two dogs, Rico and Chaser, who could retrieve large numbers of named objects; this implies not only the capacity to associate names with objects but also the ability to identify and discriminate them *as objects*. Even pigeons, it seems, have a robust concept of "pigeon," easily telling apart slides showing one or more pigeons from slides showing other species of bird![2] Of course we may parse the world differently from the way other species do, perhaps in more detail, but parsing itself must be a natural adaptation to the complex worlds we and other species inhabit.

Although the fundamental nature of thought has ancient roots, the imperative to communicate our thoughts and experiences arose much more recently. Nevertheless, it probably evolved gradually rather than in the single catastrophic event envisaged by Chomsky and others. It depended critically on a capacity especially well developed in humans but present well before the emergence of *Homo sapiens*. This was *intentionality*, a prerequisite to any communication system designed to convey variable and often novel information. Intentionality implies the capacity to voluntarily produce and understand action, and it is the primary function of the mirror system identified by Giacomo Rizzolatti and his colleagues. It is well mapped in the monkey brain and has to do primarily with the production of simple intentional acts like reaching and grasping, and in understanding such acts in others. In apes, and perhaps in some monkeys too, such actions go beyond manipulations of the physical environment to communicative acts directed to each other, generally in contexts of play and requests. Manual actions probably provided a better platform for intentional communication about events in the world than vocalization, first because they provide a natural means to convey space-time information, and second because intentionality seems to have evolved earlier in the manual domain than in the vocal one. Although most animals and many birds communicate vocally,

< 191 >

the sounds they produce are largely involuntary and designed to convey information that is emotional or affiliative rather than descriptive or explanatory.

I am of course not alone in suggesting that language evolved gradually from manual gestures. The gestural theory has a long if fitful history and is championed in the present day by authors such as Michael Arbib, Michael Tomasello, and me—and by at least one non-Michael, Giacomo Rizzolatti, along with his Italian colleagues. The theory implies that the origins of language as a communication system go back at least to the manipulative primates from whom we are descended.

Another property of thought that preceded the emergence on the planet of *Homo sapiens*, though perhaps not by much, is theory of mind—the ability to understand what others believe or are thinking. Language is in many respects an extension of theory of mind, a way of reading and influencing the minds of other. One of its critical feature is that it is *underdetermined*, and we can make sense of what others say only if we are on the same mental wavelength—a theme developed at length by Thom Scott-Phillips in his book *Speaking Our Minds*, which proposes that it is precisely our capacity for theory of mind that makes human language unique. My sense is that this may be slightly overstated, since there are examples of communicative gestures among chimpanzees that seem to be what Scott-Phillips calls *ostensive-inferential*, where the chimp may need to infer from a rather loose and ambiguous gesture what the signaler intends. But the true nature of animal communication, especially among primates, remains elusive, and there is no evidence that communication among apes can have any of the complex, recursive structure of human language. This is not to say their thoughts are not complex, or even recursive, but they do not appear to have devised an efficient system for communicating such thoughts—although the bonobo Kanzi, with his use of a multisymbol keyboard, might be on the right track. Another example, described in chapter 4, might be the chimpanzee who could select a lexigram for a given food item and then point to where that item had been hidden some time earlier.

< 192 >

The drift to more complex communication probably began with the Pleistocene, dating from about 2.9 million years ago, as our hominin forebears were forced from the dwindling forest canopy, perhaps initially to coastal areas to forage on the water's edge and later to the more open savanna and grasslands of Africa. They would have been faced with new threats, including the killer cats of the African savanna, and possibly waterborne threats as well, such as the stingrays and jellyfish that inhabit some of the waters around here. Sharing of information and experience would have been especially critical in adapting to these new environments; through language, individuals could take on the knowledge and experiences of other individuals and even of the group as a whole. They sought survival and safety in numbers, establishing the so-called cognitive niche, with theory of mind and language itself emerging to enhance cooperation and information exchange within the group. The eventual development of grammatical language was probably part of this process, founded not only on spatial understanding but also on enhanced ability to remember episodes, plan futures, and read the minds of others. These additions and embellishments are perhaps best indexed by the threefold increase in brain size through the Pleistocene.

If the sharing of experience was so adaptive, one might again ask why humans appear to be unique in developing such sophisticated means of accomplishing it. I suggested some answers in chapter 3, but here we might entertain the darker thought that in an evolutionary context language may well be a mixed blessing, perhaps even a poison dart. The imperative to share was forged against an earlier instinct to compete and fight—the nature "red in tooth and claw," as Alfred, Lord Tennyson put it in his poem *In Memoriam A.H.H.* Of course other species do share, but their sharing is largely between parents and offspring. Family ties remain strong in our own species as well, but as the social group expands beyond the familial, sharing and competitiveness vie for dominance in politics as in other realms of corporate life. In team sports, for example, cooperation within each team is paramount, although of course opposing teams compete.[3] And then there are individual sports, where nice guys finish first.[4] The

< 193 >

establishment of strong communicative bonds within groups operates to exclude individuals from other groups, creating antagonism, misunderstanding, and conflict between groups, a tendency that has escalated into global significance as group allegiances have expanded on the international stage. The six thousand languages of the world are a testimony at once to the cohesiveness and the divisiveness of the worlds we inhabit. To be sure, many languages are disappearing, but the coalescence into fewer, larger, and more powerful groupings may signal increased danger to survival. Paradoxically, then, sharing increases the size and effectiveness of competing groups and makes their conflicts all the more dangerous. Language and its offshoots may bring about the extinction of us all.

Of course the sharing of experience is not exclusively dependent on language. Families share through cohabitation, and language is often unnecessary. More broadly, we can all share a movie, a rock concert, an opera, simply by being present in the audience. The destruction of the Twin Towers on September 11, 2001, was witnessed by millions simultaneously as it was broadcast on television screens across the world. No words were necessary—apart from expressions of shock and dismay, and these are not so much linguistic as emotional. The power of language is that it can take us away from present experience and allow us to share experiences in other places and other times, and with people who were not present at the time. This need not mean, though, that other species do not relive past experiences or imagine future ones—it's just that they appear to have little ability or desire to share them.

Early in the Pleistocene, the era during which humanity was born, communication was probably achieved through mime, in extension of the interactions now witnessed in great apes at play. Mime could be shaped to correspond to stories of hunting expeditions or plans for future forays, in which sequences of action are conveyed iconically. What you get is what you see. Over time, though, communication would have been increasingly standardized, or conventionalized, ensuring that individuals in the group would come to interpret the same mimed gestures in the same way. In this process, the iconic compo-

< 194 >

nent could be increasingly relaxed and replaced by forms selected more for speed and efficiency. It was probably expedient for vocal elements to be introduced, with the signal increasingly miniaturized from movement of the body and hands to movements of the face and then to movements largely contained within the mouth. These movements could then be distinguished from each other by the production of sound, modulated into different patterns by the movements of the tongue, lips, velum, and larynx. This gradual swallowing of gesture into the mouth is not complete, because speech is still accompanied by movements of the hands and external parts of the face, and even spoken words have an iconic component. But in speech the voiced signal is both necessary and sufficient, as demonstrated by radio and telephone systems. Speech is a triumph of miniaturization.

Speech can therefore be seen as an endpoint in the shift to conventionalization and increased efficiency, although even spoken words tend to be calibrated to enhance efficiency of communication. Words (and signs) become shortened and simplified as their frequency increases. We have gone from *university* to *varsity* to *uni*, and *uniform resource locator* didn't stand a chance against *URL*. The relation between frequency and word length is most evident in the way words are written: three-letter words are the most frequent, with word frequency falling dramatically on either side.[5] The calibration of word length by frequency is an example of the principle of least effort.[6]

Speech may well count among the more powerful inventions of our species, perhaps along with the wheel, the telescope, the internal combustion engine, the computer, the Internet, the iPhone. Speech is the most efficient form of bodily language, using little energy and being available at night or when there is no visual contact with the hearer. It largely freed the rest of the body for other functions—making things, carrying things, perhaps even walking (although I did once have a colleague who was unable to walk and talk at the same time). My guess is that it was the invention of speech, not of language itself, that led to the so-called great leap forward described in chapter 2—although just how much of a leap it was remains contentious. It may have been speech that enabled us to become modern.[7]

< 195 >

But perhaps the most important contribution of language is not simply that it allows the sharing of experience, knowledge, and plans but that it allows the telling of stories and the creation of culture. As extensions of play and the imaginative construction of experience, and indeed of large chunks of our own lives, stories—that is, the need to tell them—may have provided the primary impetus for the evolution of language itself. But we do not tell all. I suspect, in fact, that we all harbor a great many experiences and chunks of knowledge that we do not share, and indeed prefer not to share, or perhaps don't even have the power to share. As we have seen language, for all its sophistication, is underdetermined, relying more on powers of inference than on what is actually available in the message. As humans, we do retain the luxury of privacy.

Many have regarded abstraction as a key attribute of language. But the use of abstract symbols is not so much a necessary property of language itself as a result of conventionalization and the drive to more economical forms of expression. Even so, the emergence of abstract symbols stripped of iconic content may well have provided extra computational power and influenced our ways of thinking. Mathematics is an extreme example, for there abstraction has proceeded to the point that a single letter, x or y, can stand for variables of wide and general reference. Linguistic analysis often proceeds at this level and can provide genuine insights into how language works. But the invention of abstract symbols probably was not a result of any big bang in our evolutionary history but was itself a product of gradual evolution. Remember that William James (quoted in chapter 6) noted two kinds of language, the discursive language that underlies storytelling and the form of language used for reasoning. Symbolic language may well have enhanced the human ability to reason, but the real roots of language probably belong in the realms of narrative.

Although I have argued that language emerged out of the common ways in which we parse our surroundings and record events, we should not overlook the extraordinary diversity among the six thousand or so languages of the world. They differ most obviously in that they use different words, but even simple translation word to word

< 196 >

seldom works. This is partly because languages order words differently and use different systems of inflections and word combinations to signal time and order of events. This can make translation notoriously difficult. A well-known but probably apocryphal example is that the biblical sentence "The spirit is willing but the flesh is weak" is said to have been translated into Russian into the equivalent of "The vodka is excellent but the meat is rotten." Even that isn't quite right, because Russian has no articles; according to Google Translate, it would actually come out as "Водка отлично, но мясо гнило"— five words instead of nine. This illustrates the further problem that languages can differ quite markedly in grammatical structure, as we saw in chapter 1, even though they share the goal of referring to experiences that have a common basis across cultures. One reason for this is that language is in part a secret code, allowing communication within groups but preventing it between groups. This is useful in on-field planning that the enemy can't follow. In World War II, bilingual American marines in the Pacific theater used the Navajo language to transmit coded messages, in the sure expectation that the enemy would not be able to understand them. This is known as code talking and had been pioneered in World War I by Cherokee and Choctaw Indians. But it goes beyond warfare. Even teenagers alter their language to the point that it is often intelligible to their parents.

Language is also creative, in the same way that evolution itself is. Languages mutate, and some mutations are selected because they prove expedient for a given culture or context. In the course of time, this has resulted in languages drifting apart, first into dialects and then into languages that are mutually unintelligible. It remains something of an open question, though, whether all current languages derive from a single source, the elusive Proto-World, or whether they have simply emerged from independently from small groupings, even single families, and then coalesced before splitting again. Accounts of the origins of language diversity typically ignore sign languages, which do seem to emerge independently and largely autonomously.

Viewed as a tool, language is also an example of human ingenuity—an ingenuity evident in mechanical tools as much as in languages themselves. We see this not only in the structure of language itself but also in the various ways in which it is manifest physically—gesture, speech, writing, the printing press, the keyboard—and each of these influences the structure of language itself. Twitter expresses things very differently from a novel by Henry James.

Social life itself would have imposed demands on language as individuals assumed different roles and social structure grew more complex. People would have assumed different roles in their communities, which would itself have required more explicit communication. "Language," writes the Hungarian biologist Eörs Szathmàry, "allows for something unprecedented: negotiated division of labor. Just as the evolution of powerful genetic and epigenetic inheritance systems allowed the evolution of complex multicellularity, natural language allowed the emergence of complex human societies."[8] And that in itself has influenced the structure of language—lawyers don't talk like people on the farm (where I grew up), or even like engineers.

Although linguists have generally focused on the sentence as the fundamental challenge for the understanding of grammar, language is much more than a string of sentences. We use language to tell stories, to share internal mental narratives. In order to tell a story we must first imagine it. In the simplest case, the structure of a story mimics the structure of the narrative; we describe events in the order in which they happened. Again, though, there is flexibility. A story will sometimes go back to earlier episodes or insert explanatory sections—much as a mathematical proof will insert lemmas to bridge gaps in logical argument. These insertions and interruptions may occur in the internal narrative itself, as we break the imagined sequence of events to consider some associated happening, or reflect on the relationship between characters, and then pick up the thread again. Or they may occur in the translation from narrative to story, perhaps to help the listener or reader to process the narrative more easily. Stories seem utterly critical to the human condition, whether oral tales of heroes or explorers, or the crime fiction and romantic sagas that haunt our reading and our TV screens. As Brian Boyd re-

< 198 >

minded us, religions are based much more on stories than on rational argument.

The ability to create narratives may well be uniquely human, independently of the language we use to share them. But again the difference is one of degree, not of kind—narratives are successions of episodes. Stories involve structures analogous to those of the sentence but at a higher level. This includes recursion—the embedding of subnarratives within the main one; indeed language itself, from words to stories, is massively hierarchical and plausibly a product of unbounded Merge. Chomsky and the generative linguists cannot be denied insights into how language actually works, although even operations like Merge may derive from simple actions like grasping an object, as suggested by Alistair Knott. And we saw earlier in this book, in chapter 4, how different regions of the hippocampus may sustain narratives at different scales, suggesting recursive embedding.

This scenario, as sketched above and elaborated in the previous chapters, seems to me much more plausible biologically than the notion of a sudden "big bang" that transformed the human mind at some point within the past one hundred thousand years, as some modern pundits proclaim. It must be said that human language remains an extraordinary phenomenon, unparalleled in nature, but by the same token we should allow millions of years, rather than the mere moment of a single mutation, to account for its evolution. We can see something of its origins in chimpanzee gesture, in the bonobo Kanzi's use of a keyboard resembling a much-enlarged iPad, and even in a monkey's grasp—not to mention in the fluency of sign languages themselves. It is unfortunate that we seem to have eradicated all of the other hominin species that inhabited the earth for millions of years since we separated from the other great apes, and that we are in other ways blind to most of what happened during that period—although archaeology and genetics do help fill some of the gap. But we must surely conclude that language did evolve in stepwise fashion, buttressed by the gradual emergence of a complex social mind, and indeed contributing to it.

If not a giant leap for humankind, then, expressive language was

< 199 >

at least a significant series of steps—or swimming strokes, if we're to persist with the Rubicon analogy. I should now confess, though, that the analogy can be misleading in that it implies a goal—crossing a river. It would of course be wrong to suggest that language was a goal to be achieved. Rather, it no doubt evolved through a series of steps, each the outcome of natural selection. Nevertheless, however achieved, language is perhaps comparable to another emergent capability, alas, denied to humans, but alike in terms of distance from its predecessors. I refer to flight—which might have enabled a crossing of the Rubicon, albeit in a more literal sense. Indeed I wonder if language is to animal communication as flight is to jumping or hopping. Whether language was initially gestural, as I have argued, or vocal, its power in allowing the sharing of thought is immense—and unprecedented, as far as we know, in the animal kingdom. People are capable of complex thoughts, whether in the form of remembered or planned episodes, or of theories about the world, or of wondrous stories, and language allows these to be shared. This sharing not only increases the repertoires of individuals but also creates culture, the common knowledge, beliefs, expertise, expectations, and prejudices that people share. As I suggested earlier, it may have been the telling of stories that provided the most powerful selective pressure toward language.

Grammar may owe some aspects of its evolution to the manufacture of tools, especially as they began to take on combinatorial properties. But the influence went the other way as well. What is especially remarkable is the spurt of sophisticated technology, along with other creative enterprises, that emerged in our own species, *Homo sapiens*, eventually setting us apart from other hominin species and perhaps even leading to the extinction of our onetime companions the Neanderthals and Denisovans—although we are increasingly led to wonder whether they were truly species distinct from our own. The rise in technological sophistication seems to have been especially pronounced within the past one hundred years, some hundred thousand or more years after our species is thought to have actually emerged. John F. Hoffecker, whom I also quoted in chapter 2, writes:

< 200 >

Modern human technological ability seems to be an integral part of a wider package of behavior ("behavioral modernity") that developed in the context of the African MSA [Middle Stone Age] before 50 ka. Modern human technology exhibits many of the characteristics, most notably the creativity and structural complexity, of art, music, ornament, and other forms of symbolism (*including by implication syntactical language*) that are elements of behavioral modernity. Modern humans, as they dispersed out of Africa, adapted quickly to a wide range of habitats designed during the late African MSA or created in response to local conditions.[9]

As suggested earlier, though, the rise of technology may have been not so much a result of symbolism as a freeing of the hands from the demands of communication itself.

Once unleashed by the freeing of the hands from communication, however, technology has undoubtedly influenced language in other ways. It has enabled people to migrate and adapt to vastly different environments, from the equator to the poles (well, almost), which accounts for the proliferation of different languages, each adapted to its culture. Although hominins had migrated from Africa beginning perhaps two million years ago, our own species, *Homo sapiens*, emerged as a species in Africa about two hundred thousnd years ago and began to migrate out of Africa perhaps around sixty thousand years ago. Presumably, as each group adapted to a new environment, it adapted both genetically and linguistically in what is called the "serial founder effect."[10] This resulted in the thousands of languages in today's world, although cultural domination is now producing a reverse effect, with many traditional languages in danger of disappearing.

Technology has also influenced sheer vocabulary size, which can range from a few thousand words in preindustrial societies to tens of thousands in the industrial world. The modern world has needed to coin new nouns such as *helicopter, vaccine, television, hard drive, Internet,* and verbs such as *download, google, tweet*—and I'm told there's a new word in Japanese, *obamu,* which means "to Obama, or persevere

< 201 >

with optimism, ignoring obstacles." The number of verbs in different languages is said to range from several thousand to thirty.[11] Of
course it is difficult to compare vocabulary sizes accurately, because
some languages cram more information into single words than other
languages do, but there seems little doubt that there are large differences in how many objects and actions can be referred to using words.

The language of the Pirahã, an Amazonian tribe whose language
has been documented by Daniel Everett, is said to have no words for
numerals and no system for counting, although a mother can easily
name and identify each one of her six children. The Pirahä have no
color words. But they can discriminate between different colors and
use metaphors to describe some colors—for example, in words loosely
translatable as "it is like blood" to describe something red, or "it is
unripe" to describe blue or green.[12] Language is adapted to culture;
we share only what we need to share.

As we have seen, the relatively small vocabularies of some languages do not mean that these languages are less complex than the
bloated languages of the Western world. Some of the most grammatically complex languages, as we have seen, are those of indigenous
peoples. There seems to have been something of a shift from highly
inflected languages, like the Australian language Walpiri, toward languages in which the grammatical components are pulled apart into
separate words: "I will plant potatoes for them again" instead of the
single Cayugan word *Ęskakhǫna'tàyęthwahs*. Of course the sheer complexity of language, and the fact that children of different cultures are
almost certainly equally capable of learning any one of them, suggests
some common basis, but as we have seen the concept of universal
grammar itself now seems without foundation, at least in a linguistic sense. No doubt children of all cultures would be equally capable
of learning to surf the Internet or fly a jetliner, but these capacities
are legacies of being human, not of universal pilotry. Everett may be
right—language is a tool like any other, differing from culture to culture but within the grasp of all children.[13]

So is the account I have given in this book simply a Just-So story?
In a sense it is inevitably so, since human evolution is a story based on

< 202 >

past events of which we have only meager evidence. But the ultimate Just-So story is the theory that language "just happened" in a single event within the past one hundred thousand years, in that "unimaginable transition" that Ian Tattersall refers to—almost expressing disbelief. I think we need a better story than that—one that deals with the successive, cumulative adaptations in our primate and hominin ancestry that eventually gave us not just language but the sophisticated world of social interaction, communication, and technology that characterizes our species today. With advances in archaeology, genetics, linguistics, neuroscience, and psychological understanding of how the mind works, there seems no longer any excuse to rely on miracles to explain how we crossed that damned Rubicon.

Now for another river to cross.

Notes

CHAPTER ONE

1. Friedrich Müller, *Lectures on the Science of Language*, 5th ed. (London: Longmans, Green, 1866), 391–92.

2. Gaston Dorren, *Lingo: A Language Spotter's Guide to Europe* (London: Profile Books, 2014).

3. http://www.ethnologue.com/country/VU/default/.

4. Christopher Moseley, *The UNESCO Atlas of the World's Languages in Danger: Content and Process* (Cambridge, UK, World Oral Literature Project, Occasional Paper 5, 2012).

5. Mark Turin, "Voices of Vanishing Worlds: Endangered Languages, Orality and Cognition," *Analise Social* 47 (2012): 846–69.

6. The comparison is not entirely straightforward, but one detailed study estimates the divergence to be 1.23 percent, implying 98.77 percent overlap; see Chimpanzee Sequencing and Analysis Consortium, "Initial Sequence of the Chimpanzee Genome and Comparison with the Human Genome," *Nature* 437 (2005): 69–87.

7. Well, maybe not. I once gave a talk on the origins of language, and a fellow in the audience claimed to know the first language spoken by humans. When I asked him how he knew, he explained that he had been observing us humans for millions of years.

8. The story of Genie and other so-called wild children is told by Michael Newton, *Savage Girls and Wild Boys: A History of Feral Children* (London: Faber and Faber, 2004).

9. Jacques Mehler et al., "A Precursor of Language Acquisition in Young Infants," *Cognition* 29 (1988): 143–78.

10. Peter G. Hepper, Glenda R. McCartney, and E. Alyson Shannon, "Lateralised Behaviour in First Trimester Human Fetuses," *Neuropsychologia* 36 (1998): 531–34.

11. Peter G. Hepper, Sara Shahidullah, and Raymond White, "Handedness in the Human Fetus," *Neuropsychologia* 29 (1991): 1101–11.

12. Cristina Colonnesi, Geert Jan J. M. Stams, Irene Koster, and Marc J. Noomb, "The Relation between Pointing and Language Development: A Meta-analysis," *Developmental Review* 30 (2010): 352–66.

13. Michael Tomasello, *The Origins of Human Communication* (Cambridge, MA: MIT Press, 2008).

14. Including ending a sentence with a preposition.

15. Noam Chomsky, "A Review of Skinner's 'Verbal Behavior,'" *Language* 35 (1959): 26–58.

< 205 >

16. Based on the number of words in a standard college dictionary, from Steven Pinker, *The Stuff of Thought* (London: Penguin Books, 2007).

17. Jorge Luis Borges, "The Library of Babel, http://maskofreason.files.wordpress.com /2011/02/the-library-of-babel-by-jorge-luis-borges.pdf, translated by J.E.I [James E. Irby] from *El jardín de senderos que se bifurcan* (1941), 8.

18. Fred Karlsson, "Constraints on Multiple Center-Embedding of Clauses," *Journal of Linguistics* 43 (2007): 365–92.

19. Nicholas Evans, *Dying Words: Endangered Languages and What They Have to Tell Us* (Oxford: Wiley-Blackwell, 2009).

20. Noam Chomsky, "Some Simple Evo Devo Theses: How True Might They Be for Language?" in *The Evolution of Human Language*, ed. Richard K. Larson, Viviane Déprez and Hiroko Yamakido (Cambridge: Cambridge University Press, 2010), 45–62.

21. Noam Chomsky, "On Cognitive Structures and Their Development: A Reply to Piaget," in *Language and Learning: The Debate between Jean Piaget and Noam Chomsky*, ed. Massimo Piattelli-Palmarini (Cambridge, MA: Harvard University Press, 1980), 48.

22. Jo-Wang Lin, "Time in a Language without Tense: The Case of Chinese," *Journal of Semantics* 23 (2006): 1–53.

23. Nicholas Evans and Stephen C. Levinson, "The Myth of Language Universals: Language Diversity and Its Importance for Cognitive Science," *Behavioral and Brain Sciences* 32 (2009): 429–92.

24. Turin, "Voices of Vanishing Worlds," 867.

25. Judy Kegl, Ann Senghas, and Marie Coppola, "Creation through Contact: Sign Language Emergence and Sign Language Change in Nicaragua," in *Language Creation and Language Change: Creolization, Diachrony, and Development*, ed. Michel DeGraff (Cambridge, MA: Bradford Book, MIT Press, 1999), 179–237.

26. Evans and Levinson, "Myth of Language Universals," 438.

27. Michael Tomasello, "Universal Grammar Is Dead," *Behavioral and Brain Sciences* 32 (2009): 470. And yet it lives on in the literature. According to the Web of Science, references to "universal grammar" in the scientific literature have risen steadily since 1965, rising to peaks of over 1,100 in 2012 and 2013, and on track to continue through 2014. A Google search produces well over two million hits.

28. The trend was documented by Thomas Corwin Mendenhall, "The Characteristic Curves of Composition," *Science* 9 (1887): 237–49, and probably needs updating. LOL.

29. Brian Hayes, "Belle Lettres Meets Big Data," *American Scientist* 102 (2014): 265.

30. Alison Wray, "Dual Processing in Protolanguage: Performance without Competence," in *The Transition to Language*, ed. Alison Wray (Oxford: Oxford University Press, 2002), 114.

31. Stephen Fry and Hugh Laurie, "Language Conversation," *A Bit of Fry and Laurie*, http://abitoffryandlaurie.co.uk/sketches/language_conversation. Some of Fry's phrases are perhaps more stock than others.

32. James R. Hurford, *The Origins of Grammar: Language in the Light of Evolution II* (Oxford: Oxford University Press, 2012).

33. Rens Bod, *Beyond Grammar: An Experience-Based Theory of Grammar* (Stanford, CA: CSLI Publications, 1998).

34. Rens Bod, "From Exemplar to Grammar: A Probabilistic Analogy-Based Model of Language Learning," *Cognitive Science* 33 (2009): 752–93.

< 206 >

35. Morten H. Christiansen and Nick Chater, "Toward a Connectionist Model of Recursion in Human Linguistic Performance," *Cognitive Science* 23 (1998): 157.

36. Adele E. Goldberg, *Constructions: A Construction Grammar Approach to Argument Structure* (Chicago: University of Chicago Press, 1995).

37. Vyvyan Evans, "Cognitive Linguistics," *WIRES Cognitive Science* 3 (2012): 129–41.

38. See, for example, Vyvyan Evans, *The Language Myth: Why Language Is Not an Instinct* (Cambridge: Cambridge University Press, 2004), or Christina Behme, "A 'Galilean' Science of Language," *Journal of Linguistics* 50 (2014): 671–704, for trenchant critiques of Chomskian theory.

39. Nicole Harlaar et al., "Predicting Individual Differences in Reading Comprehension: A Twin Study," *Annals of Dyslexia* 60 (2010): 265–88.

40. And even that famous remark seems to have arisen from the shoulders of others. In 1159 St. John of Salisbury wrote (in translation), "Bernard of Chartres used to say that we are like dwarves on the shoulders of giants." This is according to Google, and "Stand on the shoulders of giants" has become the motto of Google Scholar.

CHAPTER TWO

1. Alfred Russel Wallace, "Sir Charles Lyell on Geological Climates and the Origin of Species," *Quarterly Review* 126 (1869): 359–94.

2. Derek C. Penn, Keith J. Holyoak and Daniel J. Povinelli, "Darwin's Mistake: Explaining the Discontinuity between Human and Nonhuman Minds," *Behavioral and Brain Sciences*, 31 (2008): 108–78. Others, though, have similarly documented the "gap" but attempted to provide evolutionary accounts; see for example Thomas Suddendorf, *The Gap: The Science of What Separates Us from Other Animals* (New York: Basic Books, 2013).

3. David Premack, "Gavagai! or The Future History of the Animal Language Controversy," *Cognition* 19 (1985): 207–96.

4. Richard Dawkins, *The Blind Watchmaker: Why the Evidence of Evolution Reveals a Universe without Design* (New York: W. W. Norton, 1986).

5. Noam Chomsky, "Some Simple Evo Devo Theses: How True Might They Be for Language?" in *The Evolution of Human Language*, ed. Richard K. Larson, Viviane Déprez, and Hiroko Yamakido (Cambridge: Cambridge University Press, 2010), 58.

6. As Chomsky himself puts it, Merge is "an indispensable operation of a recursive system . . . which takes two syntactic objects A and B and forms the new object G={A,B}." From Noam Chomsky, *Derivation by Phase* (Cambridge, MA: MIT Press,1999), 2.

7. Chomsky, "Some Simple Evo Devo Theses," 59.

8. Noam Chomsky, "Biolinguistic Explorations: Design, Development, Evolution," *International Journal of Philosophical Studies* 15 (2007): 18.

9. Derek Bickerton, *More than Nature Needs: Language, Mind, and Evolution* (Cambridge, MA: Harvard University Press, 2014), 117.

10. Gravity does lots of other things too, like holding the world together, but let's not go there just now.

11. Steven Pinker, *The Stuff of Thought* (London, Penguin Books, 2007).

12. Elizabeth S. Spelke and Katherine D. Kinzler, "Core Knowledge," *Developmental Science* 10 (2007): 89–96.

13. In some accounts it was the hostess who spilled the tea, and in some the beverage was

tea, not coffee. Check Google. The British comedian Frank Muir is reported to have said that the temptation to spill coffee onto such a child must have been quite strong.

14. Chomsky, "Some Simple Evo Devo Theses," 87.

15. We shall see in chapter 5 that spoken words are not quite as arbitrary, or "noniconic," as generally supposed.

16. These and other examples are easily found on YouTube.

17. Juliane Kaminski, Josep Call, and Julia Fischer, "Word Learning in the Domestic Dog: Evidence for 'Fast Mapping,'" *Science* 304 (2004): 1682–83.

18. John W. Pilley and Alliston K. Reid, "Border Collie Comprehends Object Names as Verbal Referents," *Behavioural Processes* 86 (2011): 184–95.

19. Heidi Lyn et al., "Apes Communicate about Absent and Displaced Objects: Methodology Matters," *Animal Cognition* 17 (2014): 85–94.

20. Paul A. Mellars, "Going East: New Genetic and Archaeological Perspectives on the Modern Colonization of Eurasia," *Science* 313 (2006): 796–800.

21. Juan Luis Arsuaga et al., "Neandertal Roots: Cranial and Chronological Evidence from Sima de los Huesos," *Science* 344 (2014): 1356–73.

22. Ron Pinhasi et al., "Revised Age of Late Neanderthal Occupation and the End of the Middle Paleolithic in the Northern Caucasus," *Proceedings of the National Academy of Sciences* 108 (2011): 8611–16.

23. Richard E. Green et al., "A Draft Sequence of the Neanderthal Genome," *Science* 328 (2010): 710–22.

24. Qiaomei Fu et al., "An Early Modern Human from Romania with a Recent Neanderthal Ancestor," *Nature* 524 (2015): 216–19.

25. David Reich et al., "Genetic History of an Archaic Hominin Group from Denisova Cave in Siberia," *Nature* 468 (2010): 1053–60.

26. Richard G. Klein, "Out of Africa and the Evolution of Human Behavior," *Evolutionary Anthropology* 17 (2008): 271.

27. John F. Hoffecker, "Representation and Recursion in the Archaeological Record," *Journal of Archaeological Method and Theory* 14 (2007): 379.

28. Paul A. Mellars, "Origins of the Female Image," *Nature* 459 (2006): 177.

29. Ian Tattersall, *Masters of the Planet: The Search for Human Origins* (New York: Palgrave Macmillan, 2012), 199.

30. Terrence Deacon, *The Symbolic Species* (New York, W. W. Norton, 1997).

31. Timothy J. Crow, "Sexual Selection, Timing, and an X-Y Homologous Gene: Did *Homo sapiens* Speciate on the Y Chromosome?" in *The Speciation of Modern Homo Sapiens*, ed. Timothy J. Crow (Oxford: Oxford University Press, 2002), 197–216.

32. Sally McBrearty and Alison S. Brooks, "The Revolution That Wasn't: A New Interpretation of the Origin of Modern Human Behavior," *Journal of Human Evolution* 39 (2000): 453–563.

33. Hartmut Thieme, "Lower Palaeolithic Hunting Spears from Germany," *Nature* 385 (1997): 807–10. More recent evidence suggests that spears for throwing may go back as far as 500,000 years ago; Jane Wilkins et al., "Evidence for Early Hafted Hunting Technology," *Science* 338 (2012): 942–46.

34. John J. Shea, "*Homo sapiens* Is as *Homo sapiens* Was," *Current Anthropology* 52 (2011): 1.

< 208 >

CHAPTER THREE

1. Theodosius Dobzhansky, "Nothing in Biology Makes Sense Except in the Light of Evolution," *American Biology Teacher* 35 (1973): 125–29.

2. Charles Darwin, *On the Origin of Species by Means of Natural Selection* (London: John Murray, 1859), 158.

3. Noam Chomsky, *The Architecture of Language* (New Delhi: Oxford University Press, 2000), 4.

4. Charles Darwin, *The Descent of Man and Selection in Relation to Sex*, 2nd ed. (New York: Appleton, 1871), 53.

5. Laura A. Pettito and Pauline Marentette, "Babbling in the Manual Mode: Evidence for the Ontogeny of Language," *Science* 251 (1991): 1493–96.

6. Darwin, *On the Origin of Species*, 62.

7. Noam Chomsky, *Reflections on Language* (New York: Pantheon, 1975), 59.

8. Stephen J. Gould, "Panselectionist Pitfalls in Parker and Gibson's Model of the Evolution of Intelligence," *Behavioral and Brain Sciences* 2 (1979): 385.

9. Chomsky, *Reflections on Language*, 59.

10. Michael C. Corballis, *The Recursive Mind* (Princeton, NJ: Princeton University Press, 2011).

11. Stephen J. Gould and Niles Eldredge, "Punctuated Equilibria: The Tempo and Mode of Evolution Reconsidered," *Paleobiology* 3 (1977): 115–51.

12. Stephen J. Gould, *Time's Arrow, Time's Cycle* (Cambridge, MA: Harvard University Press, 1987), 24.

13. Steven Pinker and Paul Bloom, "Natural Language and Natural Selection," *Behavioral and Brain Sciences* 13 (1990): 707–84.

14. Ray Jackendoff, *Foundations of Language: Brain, Meaning, Grammar, Evolution* (Oxford: Oxford University Press, 2002).

15. George C. Williams, *Adaptation and Natural Selection: A Critique of Some Current Evolutionary Thought* (Princeton, NJ: Princeton University Press, 1966).

16. Don't disparage the roundworm. According to Susan Cain, it's the introverts who rule—and what is more introverted than a roundworm? See Susan Cain, *Quiet: The Power of Introverts in a World That Can't Stop Talking* (London: Penguin Books, 2013).

17. Noam Chomsky, "Biolinguistic Explorations: Design, Development, Evolution," *International Journal of Philosophical Studies* 15 (2007): 3.

18. Morten H. Christiansen and Nick Chater, "Language as Shaped by the Brain," *Behavioral and Brain Sciences* 31 (2008): 496.

19. Pinker and Bloom, "Natural Language and Natural Selection," 726.

20. Mark Verhaegen, "The Aquatic Ape Evolves: Common Misconceptions and Unproven Assumptions about the So-called Aquatic Ape Hypothesis," *Human Evolution* 28 (2013): 237–66.

21. John Tooby and Irven DeVore, "The Reconstruction of Hominid Evolution through Strategic Modeling," in *The Evolution of Human Behavior: Primate Models*, ed. W. G. Kinzey (Albany: SUNY Press, 1987), 183–237.

22. Sue Savage-Rumbaugh, Stuart G. Shanker, and Talbot J. Taylor, *Apes, Language, and the Human Mind* (New York: Oxford University Press, 1998).

23. James R. Hurford, *The Origins of Meaning: Language in the Light of Evolution* (Oxford:

< 209 >

Oxford University Press, 2007), and *The Origins of Grammar: Language in the Light of Evolution II* (Oxford: Oxford University Press, 2012).

24. W. Tecumseh Fitch, *The Evolution of Language* (Cambridge: Cambridge University Press, 2010).

25. To some extent, Chomsky himself was complicit in making this distinction, since he was a coauthor in the publication that first made it; see Marc D. Hauser, Noam Chomsky, and W. Tecumseh Fitch, "The Faculty of Language: What Is It, Who Has It, and How Did It Evolve?" *Science* 298 (2010): 1569–79.

26. Fitch, *Evolution of Language*, 118.

27. Murray Gell-Mann and Merritt Ruhlen, "The Origin and Evolution of Word Order," *Proceedings of the National Academy of Sciences USA* 108 (2011): 17,290–95.

28. Jared Diamond, "Deep Relationships among Languages," *Nature* 476 (2011): 291–92.

29. Nicholas Evans, *Dying Words: Endangered Languages and What They Have to Tell Us* (Oxford: Wiley-Blackwell, 2009).

30. Daniel L. Everett, "Cultural Constraints on Grammar and Cognition in Pirahã," *Current Anthropology* 46 (2005): 621–46.

31. Manuel Carreiras et al., "Neural Processing of a Whistled Language," *Nature* 433 (2005): 31–32.

32. Ann Salmond, "Mana Makes the Man: A Look at Maori Oratory and Politics," in *Political Language and Oratory in Traditional Society*, ed. M. Bloch (New York: Academic, 1975), 50.

33. Quentin D. Atkinson, "Phonemic Diversity Supports a Serial Founder Effect Model of Language Expansion from Africa," *Science* 332 (2011): 348. Atkinson's conclusions have been queried, though, by Michael Cysouw, Dan Dediu, and Steven Moran ("Comment on 'Phonemic Diversity Supports a Serial Founder Effect Model of Language Expansion from Africa,'" *Science* 332 [2012]: 346–48), who suggest that Atkinson's data are more consistent with language origins deriving from at least two locations, one in eastern Africa and one in the Caucasus. Using more refined analyses, moreover, they identify several other "founder effects." The search for the origins of language remains fraught.

34. M. Paul Lewis, Gary F. Simons, and Charles D. Fennig, eds., "Vanuatu," in *Ethnologue: Languages of the World*, 19th ed. (Dallas, Texas: SIL International, 2016), http://www.ethnologue.com/country/VU/default/.

35. Daniel Dor, *The Instruction of Imagination: Language as a Social Communication Technology* (New York: Oxford University Press, 2015), 12.

36. Daniel L. Everett, *Language: The Cultural Tool* (New York: Pantheon, 2012), 6.

37. Dor, *Instruction of Imagination*, 12.

CHAPTER FOUR

1. John Locke, *An Essay concerning Human Understanding* (orig. 1690; London: Scolar, 1970), bk. 3, chap. 10.

2. Plato, *The Dialogues of Plato*, trans. and ed. B. Jowett, vol. 4 (Oxford: Oxford University Press, 1892).

3. Immanuel Kant, "Anthropologie," in *Gesammelte Schriften* (1798), trans. in "The Grammar of Reason: Hammann's challenge to Kant," *Synthèse* 75 (1988): 278.

4. Jerry A. Fodor, *The Language of Thought* (New York: Crowell, 1975).

5. Steven Pinker, *The Stuff of Thought* (London: Penguin Books, 2007).

< 210 >

6. Merlin Donald, *Origins of the Modern Mind* (Cambridge, MA: Harvard University Press, 1991), 253.

7. Sadly, Turing took his own life in 1954 after being convicted of homosexual acts, then regarded as criminal, and subjected to treatment with female hormones (chemical castration). Ironically, the original Turing test was a game in which a man and a woman are sent into separate rooms and guests try to tell them apart by writing a series of questions and reading the typewritten answers sent back. The man and the woman each try to pretend to be the other.

8. Alan Newell, John Calman Shaw, and Herbert A. Simon, "Elements of a Theory of Human Problem Solving," *Psychological Review* 23 (1958): 151–66.

9. Zenon W. Pylyshyn, "What the Mind's Eye Tells the Mind's Brain: A Critique of Mental Imagery," *Psychological Bulletin* 80 (1973): 1–24.

10. Giorgio Ganis and Haline E. Schendan, "Visual Imagery," *WIRES Cognitive Science* 2 (2011): 239–52.

11. Ray Jackendoff, "What Is the Human Language Faculty? Two Views," *Language* 87 (2011): 586–624.

12. Noam Chomsky, *New Horizons in the Study of Language and Mind* (Cambridge: Cambridge University Press, 2000).

13. Steven Pinker, *The Stuff of Thought* (London: Penguin Books, 2007).

14. Daniel Dor, *The Instruction of Imagination: Language as a Social Communication Technology* (New York: Oxford University Press, 2015), 1.

15. Lev Vygotsky, *Mind in Society* (Cambridge, MA: Harvard University Press, 1978).

16. Steven Pinker and Paul Bloom, "Natural Language and Natural Selection," *Behavioral and Brain Sciences* 13 (1990): 482.

17. Katherine Nelson, *Language in Cognitive Development: The Emergence of the Mediated Mind* (Cambridge: Cambridge University Press, 1996).

18. Dan I. Slobin, "From 'Thought and Language' to 'Thinking for Speaking,'" in *Rethinking Linguistic Relativity*, ed. John Gumperz and Stephen C. Levinson (Cambridge: Cambridge University Press, 1960), 70–96.

19. Randy L. Buckner, Jessica R. Andrews-Hanna, and Daniel L. Schacter, "The Brain's Default Network: Anatomy, Function, and Relevance to Disease," *Annals of the New York Academy of Sciences* 1124 (2008): 1–38.

20. Michael Corballis, *The Wandering Mind* (Chicago: University of Chicago Press, 2015).

21. Marcus E. Raichle et al., "A Default Mode of Brain Function," *Proceedings of the National Academy of Sciences* 109 (2001): 3979–84.

22. First published in the *New Yorker* in 1939 and made into a film starring Danny Kaye in 1947. Another version, starring Ben Stiller, appeared in 2013.

23. I did once dream I was being attacked by a horse, but finding myself unable to escape, I meekly allowed it to eat me.

24. J. Allan Hobson, "REM Sleep and Dreaming: Towards a Theory of Protoconsciousness," *Nature Reviews Neuroscience* 10 (2009): 803–14.

25. Carlo Rovelli, *Seven Brief Lessons on Physics* (London: Penguin Books, 2015) 1.

26. Juergen Fell, "I Think, Therefore I Am (Unhappy)," *Frontiers in Human Neuroscience* 6 (2012), art. 132.

< 211 >

27. Endel Tulving, "Episodic and Semantic Memory," in *Organization of Memory*, ed. Endel Tulving and Wayne Donaldson (New York: Academic, 1972), 381–403.

28. Stanley B. Klein, "The Complex Act of Projecting Oneself into the Future," *WIREs Cognitive Science* 4 (2013): 63–79.

29. From his poem "My Future," published in the *Spectator* of July 26, 2014, and reprinted here with his permission.

30. Elizabeth Loftus and Katherine Ketcham, *The Myth of Repressed Memory* (New York: St. Martin's, 1994).

31. Alan C. Kamil and Russell P. Balda, "Cache Recovery and Spatial Memory in Clark's Nutcrackers (*Nucifraga columbiana*)," *Journal of Experimental Psychology: Animal Behavior Processes* 85 (1985): 95–111.

32. Endel Tulving, "Memory and Consciousness," *Canadian Psychologist* 26 (1985): 1–12.

33. Thomas Suddendorf and Michael C. Corballis, "Mental Time Travel and the Evolution of the Human Mind," *Genetic, Social, and General Psychology Monographs* 123 (1997): 133–67.

34. Stanley B. Klein, Theresa E. Robertson, and Andrew W. Delton, "Facing the Future: Memory as an Evolved System for Planning Future Acts," *Memory and Cognition* 38 (2010): 21.

35. Joseph Dien ("Looking Both Ways through Time: The Janus Model of Lateralised Cognition," *Brain and Cognition* 67 [2008]: 292–23) has also suggested that the left brain is oriented to the future and the right brain to the past, but he says that this may be characteristic of nonhuman species as well as of humans. He calls this the Janus theory of brain asymmetry.

36. Nicola S. Clayton, Timothy J. Bussey, and Anthony Dickinson, "Can Animals Recall the Past and Plan for the Future?" *Trends in Cognitive Sciences* 4 (2003): 685–91.

37. Charles R. Menzel, "Unprompted Recall and Reporting of Hidden Objects by a Chimpanzee (*Pan Troglodytes*) after Extended Delays," *Journal of Comparative Psychology* 113 (1969): 426–34.

38. Donald R. Griffin, *Animal Minds: From Cognition to Consciousness* (Chicago: University of Chicago Press, 2001), 1.

39. Anne Russon and Kristin Andrews, "Orangutan Pantomime: Elaborating the Message," *Biology Letters* 7 (2001): 627–30.

40. Fumihiro Kano and Satoshi Hirata, "Great Apes Make Anticipatory Looks Based on Long-Term Memory of Single Events," *Current Biology* 25 (2015): 1–5.

41. Sergio P. C. Correia, Anthony Dickinson, and Nicola S. Clayton, "Western Scrub-Jays Anticipate Future Needs Independently of Their Current Motivational State," *Current Biology* 17 (2007): 856–61.

42. Nicholas J. Mulcahy and Josep Call, "Apes Save Tools for Future Use," *Science* 312 (2006): 1038–40.

43. Mathias Osvath and Elin Karvonen, "Spontaneous Innovation for Future Deception in a Male Chimpanzee," *PLoS ONE* 7 (2012), e36782.

44. Charles Kingsley, *The Water-Babies: A Fairy Tale for a Land Baby* (London: Macmillan, 1863), 8.

45. Quoted by Charles G. Gross, "Huxley versus Owen: The Hippocampus Minor and Evolution," *Trends in Neuroscience* 16 (1993): 497. The poem was signed "Gorilla" and written by Sir Philip Egerton, an English MP and paleontologist. It is a rather plodding example of *Punch* humor.

46. Charles G. Gross, "Huxley versus Owen."

< 212 >

47. Donna Rose Addis, Alana T. Wong, and Daniel L. Schacter, "Remembering the Past and Imagining the Future: Common and Distinct Neural Substrates during Event Construction and Elaboration," *Neuropsychologia* 45 (2007): 1363–77.

48. Demis Hassabis, Dharshan Kumaran, and Eleanor A. Maguire, "Using Imagination to Understand the Neural Basis of Episodic Memory," *Journal of Neuroscience* 27 (2007): 14,365–74.

49. Justin L. Vincent et al., "Intrinsic Functional Architecture in the Anaesthetized Monkey Brain," *Nature* 447 (2007): 83–86.

50. Hanbing Lu et al., "Rat Brains Also Have a Default Mode Network," *Proceedings of the National Academy of Sciences (USA)* 109 (2012): 3979–84.

51. John O'Keefe and Lynn Nadel, *The Hippocampus as a Cognitive Map* (Oxford: Clarendon, 1978). John O'Keefe was awarded a share of the Nobel Prize in Physiology of Medicine in 2014 for this work.

52. Eleanor A. Maguire et al., "Navigation-Related Cells: Structural Change in the Hippocampi of Taxi Drivers," *Proceedings of the National Academy of Sciences (USA)* 97 (2000): 4398–403.

53. Eleanor A. Maguire, Katherine Woollett, and Hugo J. Spiers, "London Taxi Drivers and Bus Drivers: A Structural MRI and Neuropsychological Analysis," *Hippocampus* 16 (2006): 1091–101.

54. Charlotte B. Almea et al., "Place Cells in the Hippocampus: Eleven Maps for Eleven Rooms," *Proceedings of the National Academy of Sciences (USA)* 111 (2014): 18,428–35. Almea's coauthors Edvard and May-Britt Moser jointly shared the 2014 Nobel Prize with John O'Keefe.

55. Or, to be more accurate, two, since it is a bilateral structure with each lobe resembling a seahorse.

56. Anoopum Gupta et al., "Hippocampal Replay Is Not a Simple Function of Experience," *Neuron* 65 (2010): 695–705.

57. Dori Derdikman and May-Britt Moser, "A Dual Role for Hippocampal Relay," *Neuron* 65 (2010), 582–84.

58. Eva Pastalkova, Vladimir A. Itskov, and György Buzsáki, "Internally Generated Cell Assembly Sequences in the Rat Hippocampus," *Science* 321 (2008): 1322–27.

59. May-Britt Moser, David C. Rowland, and Edvard I. Moser, "Place Cells, Grid Cells, and Memory." *Cold Spring Harbor Perspectives in Biology* 7 (2015): a021808, 6.

60. Jonathan F. Miller et al., "Neural Activity in Human Hippocampal Formation Reveals the Spatial Context of Retrieved Memories," *Science* 342 (2013): 1111–14.

61. Bryan A. Strange et al., "Functional Organization of the Hippocampal Longitudinal Axis," *Nature Reviews Neuroscience* 15 (2014): 655–69.

62. Silvy H. P. Collin, Branka Milivojevic, and Christian F. Doeller, "Memory Hierarchies Map onto the Hippocampal Long Axis in Humans," *Nature Neuroscience*, 2015, advance online publication, doi:10.1038/nn.4138.

63. Moser, Rowland, and Moser, "Place Cells, Grid Cells, and Memory," 11.

64. Charles Darwin, *The Descent of Man and Selection in Relation to Sex*, 2nd ed, (New York: Appleton, 1871), 126.

65. Derek Bickerton, *More than Nature Needs: Language, Mind, and Evolution* (Cambridge, MA: Harvard University Press, 2014), 93. See also Peter Gärdenfors and Mathias Osvath,

< 213 >

"Prospection as a Cognitive Precursor to Symbolic Communication," in *The Evolution of Human Language*, ed. Richard K. Larson, Viviane Déprez, and Hiroko Yamakido (Cambridge: Cambridge University Press, 2010), 103–14.

CHAPTER FIVE

1. H. Paul Grice, *Studies in the Ways of Words* (Cambridge: Cambridge University Press, 1989), 30–31.

2. Gilles Fauconnier, "Cognitive Linguistics," in *Encyclopedia of Cognitive Science*, ed. Lynn Nadel (London: Nature, 2003), 1:540.

3. Dan Sperber and Deirdre Wilson, "Pragmatics, Modularity and Mind-Reading," *Mind and Language* 17 (2002): 15.

4. This is not intended as a slight on *Time* magazine. Wherever you go in Australia, you seem to be constantly accompanied by flies, although I'm told that this has been partly controlled by the importation of dung beetles. (I don't always believe what I'm told).

5. Thomas C. Scott-Phillips, *Speaking Our Minds: Why Human Communication Is Different, and How Language Evolved to Make It Special* (Basingstoke, UK: Palgrave Macmillan, 2015).

6. Scott-Phillips, recognizing that the phrase is cumbersome, shortens it to *ostensive communication*—itself an act of ostensive communication.

7. Charles Darwin, *The Expression of the Emotions in Man and Animals* (London: John Murray, 1872), 75–76.

8. Frans B. M. De Waal, "The Antiquity of Empathy," *Science* 336 (2012): 874–76.

9. Dale J. Langford et al., "Social Modulation of Pain as Evidence for Empathy in Mice, "*Science* 312 (2006): 1967–70.

10. Stanley Wechkin, Jules H. Masserman, and William Terris, "Shock to a Conspecific as an Aversive Stimulus," *Psychonomic Science* 1 (1964): 47–48.

11. Victoria Horner et al., "Spontaneous Prosocial Choice by Chimpanzees," *Proceedings of the National Academy of Sciences USA* 108 (2011): 13,847–51.

12. More accurately, their pupils dilate when they are shown scenes of people being helped after they drop a can when stacking cans into a tower, or drop a crayon when drawing a picture. Pupil dilation is an indicator of emotional reaction. Robert Hepach, Amrisha Vaish, and Michael Tomasello, "Young Children Are Intrinsically Motivated to See Others Helped," *Psychological Science* 23 (2012): 967–72.

13. David Premack and Guy Woodruff, "Does the Chimpanzee Have a Theory of Mind?," *Behavioral and Brain Sciences* 4 (1978): 515–26.

14. Daniel J. Povinelli, Jesse M. Bering, and Steve Giambrone, "Toward a Science of Other Minds: Escaping the Argument by Analogy," *Cognitive Science* 24 (2000): 509–41.

15. Brian Hare, Josep Call, and Michael Tomasello, "Chimpanzees Deceive a Human Competitor by Hiding," *Cognition* 101 (2006): 495–514.

16. Brian Hare et al., "Chimpanzees Know What Conspecifics Do and Do Not See," *Animal Behaviour* 59 (2000): 771–85.

17. Judith Maria Burkart and Adolf Heschl, "Perspective Taking or Behaviour Reading? Understanding of Visual Access in Common Marmosets (*Calithrix jacchus*)," *Animal Behaviour* 73 (2007): 457–69.

18. Brian Hare, Josep Call, and Michael Tomasello, "Do Chimpanzees Know What Conspecifics Know?," *Animal Behaviour* 61 (2001): 139–51.

< 214 >

19. Derek C. Penn, Keith J. Holyoak, and Daniel J. Povinelli, "Darwin's Mistake: Explaining the Discontinuity between Human and Nonhuman Minds," *Behavioral and Brain Sciences* 31 (2008): 108–78.

20. Josep Call and Michael Tomasello, "Does the Chimpanzee Have a Theory of Mind? 30 Years Later," *Trends in Cognitive Science* 12 (2008): 187–92.

21. Katie E. Slocombe and Klaus Zuberbühler, "Chimpanzees Modify Recruitment Screams as a Function of Audience Composition," *Proceedings of the National Academy of Sciences (USA)* 104 (2007): 17,228–33.

22. Thure E. Cerling et al., "Woody Cover and Hominin Environments in the Past 6 Million Years," *Nature* 476 (2011): 51–56.

23. Robert A. Foley, "Early Man and the Red Queen: Tropical African Community Evolution and Hominid Adaptation," in *Hominid Evolution and Community Ecology: Prehistoric Human Adaptation in Biological Perspective*, ed. Robert A. Foley (London: Academic, 1984), 85–110.

24. Robin Dunbar, "The Social Brain: Mind, Language and Society in Evolutionary Perspective," *Annual Review of Anthropology* 32 (2003): 163–81.

25. David G. Rand, Joshua D. Greene, and Martin A. Nowak, "Spontaneous Giving and Calculated Greed," *Nature* 489 (2012): 427–30.

26. Richard D. Alexander, *How Did Humans Evolve? Reflections on the Uniquely Unique Species*, Museum of Zoology Special Publication 1 (Ann Arbor: University of Michigan, 1990).

27. Nicholas Humphrey, "The Social Function of Intellect," in *Growing Points in Ethology*, ed. Patrick P. G. Bateson and Robert A. Hinde (Cambridge: Cambridge University Press, 1976), 311.

28. Kim Sterelny, "Social Intelligence, Human Intelligence, and Niche Construction," *Philosophical Transactions of the Royal Society B* 362 (2007): 719–30.

29. David Premack, "Human and Animal Cognition: Continuity and Discontinuity," *Proceedings of the National Academy of Sciences (USA)* 104 (2007): 13,864.

30. From a lecture delivered by Brian Boyd in August 2001. I thank Brian for sending me this extract.

31. Megan Y. Dennis et al., "Evolution of Human-Specific Neural SRGAP2 Genes by Incomplete Segmental Duplication," *Cell* 149 (2012): 1–11. Mutations in other genes have also been linked to the increase in brain size. These include the ASPM, MCPH6, and MYH16.

32. Philip V. Tobias, "Revisiting Water and Hominin Evolution," in *Was Man More Aquatic in the Past? Fifty Years after Alister Hardy*, ed. Mario Vaneechoutte, Algis Kuliakis, and Marc Verhaegen (Oak Park, IL: Bentham Science, 2011), 3.

33. Mark Verhaegen, "The Aquatic Ape Evolves: Common Misconceptions and Unproven Assumptions about the So-Called Aquatic Ape Hypothesis," *Human Evolution* 28 (2013): 237–66.

34. Missing, one might say, presumed drowned.

35. Mark Verhaegen, "Origin of Hominid Bipedalism," *Nature* 325 (1987): 305.

36. Michael A. Crawford and C. Leigh Broadhurst, "The Role of Docosahexaenoic and the Marine Food Web as Determinants of Evolution and Hominid Brain Development: The Challenge for Human Sustainability," *Nutritional Health* 21 (2012): 17–39.

37. Marcos M. Duarte, "Red Ochre and Shells: Clues to Human Evolution," *Trends in Ecology and Evolution* 29 (2014): 564.

38. John L. Locke and Barry Bogin, "Language and Life History: A New Perspective

< 215 >

on the Development and Evolution of Human Language," *Behavioral and Brain Sciences* 29 (2006): 259–325.

39. A. A. Milne, *Now We Are Six* (London: Methuen, 1927).

40. David J. Miller et al., "Prolonged Myelination in Human Neocortical Evolution," *Proceedings of the National Academy of Sciences (USA)* 109 (2012): 16,480–85.

CHAPTER SIX

1. William James, cited by Jerome Bruner, *Actual Minds, Possible Worlds* (Cambridge, MA: Harvard University Press, 1986).

2. Katherinc Nelson, *Language in Cognitive Development: The Emergence of the Mediated Mind* (Cambridge, UK, Cambridge University Press, 1996) 181.

3. Yuval N. Harari, *Sapiens: A Brief History of Humankind* (London: Harvill Secker, 2014), 24.

4. Aristotle, *Poetics*, trans. Gerald Frank Else (Ann Arbor: University of Michigan Press, 1970).

5. Keith Oatley, "The Cognitive Science of Fiction," *WIREs Cognitive Science* 3 (2012): 427.

6. Raymond A. Mar et al., "Bookworms versus Nerds: Exposure to Fiction versus Nonfiction, Divergent Associations with Social Ability, and the Simulation of Fictional Social Worlds," *Journal of Research in Personality* 40 (2006): 694–712.

7. R. Nathan Spreng, Raymond A. Mar, and Alice S. N. Kim., "The Common Neural Basis of Autobiographical Memory, Prospection, Navigation, Theory of Mind, and the Default Mode: A Quantitative Meta-analysis," *Journal of Cognitive Neuroscience* 21 (2009): 489–510.

8. Lev Vygotsky, *Mind in Society* (Cambridge, MA: Harvard University Press, 1978), 94.

9. Ibid., 99.

10. Polly W. Wiessner, "Embers of Society: Firelight Talk among the Ju/'huansi Bushmen," *Proceedings of the National Academy of Sciences* 111 (2014): 14,027–35.

11. Emiko Ohnuki-Tierney, *Illness and Healing among the Sahkalin Ainu: A Symbolic Interpretation.* (Cambridge: Cambridge University Press Archive, 1981), cited in Wiessner, "Embers of Society," 14,032–33.

12. John A. Gowlett and Richard W. Wrangham, "Earliest Fire in Africa: Towards the Convergence of Archeological Evidence and the Cooking Hypothesis," *Azania* 48 (2013): 5–30.

13. Robin Dunbar and John A. Gowlett, "Fireside Chat: The Impact of Fire on Hominin Socioecology," in *Lucy to Language: The Benchmark Papers*, edited by Robin Dunbar, Clive Gamble, and John Gowlett (Oxford: Oxford University Press, 2014), 277–96.

14. Ian D. Goodwin, Stuart A. Browning, and Athol J. Anderson, "Climate Windows for Polynesian Voyaging to New Zealand and Easter Island," *Proceedings of the National Academy of Sciences* 111:14,716–21.

15. George Grey, *Polynesian Mythology and Ancient Traditional History of the New Zealand Race* (London: John Murray, 1855).

16. John Sutherland, *A Little History of Literature* (New Haven, CT: Yale University Press, 2013).

17. *The Epic of Gilgamesh: The Babylonian Epic Poem and Other Texts in Akkadian and Sumerian* (London: Penguin, 2000).

18. Harari, *Sapiens*.

19. Brian Boyd, *The Origin of Stories: Evolution, Cognition, and Fiction* (Cambridge, MA: Belknap Press of Harvard University Press, 2009).

< 216 >

20. Richard Sosis, "The Adaptive Value of Religious Ritual," *American Scientist* 92 (2004): 166–72.

21. Boyd, *Origin of Stories*, 206.

22. Paul K. Piff et al., "Higher Social Class Predicts Increased Unethical Behavior," *Proceedings of the National Academy of Sciences USA* 109 (2012): 4087–91.

23. Rabbits are up there too.

24. This has been debunked by Ina Wunn, "Beginning of Religion," *Numen* 47 (2000): 417–52, but if not a bear cult there seems at least to be a bear cult cult.

25. Robin Dunbar, *The Human Story* (London: Faber and Faber, 2004), 185. And on top of all that, God created the universe.

26. Charles Darwin, *The Descent of Man and Selection in Relation to Sex*, 2nd ed. (New York: Appleton, 1871), 69.

27. David S. Wilson, *Darwin's Cathedral: Evolution, Religion, and the Nature of Society* (Chicago: University of Chicago Press, 2002), 64.

28. David Lewis-Williams and David Pearce. *Inside the Neolithic Mind: Consciousness, Cosmos, and the Realm of the Gods* (London: Thames and Hudson, 2005).

29. Albert Einstein, *The World as I See It* (Secaucus, NJ: Citadel, 1979), 28.

30. Quoted by Deirdre David, *The Cambridge Companion to the Victorian Novel* (Cambridge: Cambridge University Press, 2001), 179.

31. Cited in Frederick S. Frank and Anthony Magistrale, *The Poe Encyclopedia* (Westwood, CT: Greenwood, 1997), 103.

32. Real-life detectives, as they sometimes appear on TV or in the press, seem a much more prosaic lot.

33. From an interview with the *NZ Listener* of November 3, 2012.

34. Men may prefer (visual) pornography.

35. Daisy Cummins and Julie Bindel, "Mills & Boon: 100 Years of Heaven or Hell?," *Guardian*, December 5, 2007, http://www.theguardian.com/lifeandstyle/2007/dec/05/women .fiction.

36. Of course, as a man I would not dare be caught reading a romantic novel, so I don't really know what goes on there.

37. Susan Quilliam, "'He Seized Her in His Manly Arms and Bent His Lips to Hers . . .': The Surprising Impact That Romantic Novels Have on Our Work," *Journal of Family Planning, Reproduction and Health Care* 37 (2011): 179–81.

CHAPTER SEVEN

1. Birds do it, bees do it. Australian magpies point with their beaks to indicate the location of an intruder, and evidently do so with intent; see Gisela Kaplan, "Pointing Gesture in a Bird—Merely Instrumental or a Cognitively Complex Behaviour?," *Current Zoology* 57 (2011): 453–67. The waggle dance of the bees to indicate the distance and direction in which nectar may be found has been known since pioneering studies of Karl von Frisch, *The Dance Language and Orientation of Bees* (Cambridge, MA: Harvard University Press, 1967).

2. Étienne Bonnot de Condillac, *An Essay on the Origin of Human Knowledge: Being a Supplement to Mr. Locke's "Essay on the Human Understanding"* [1747], trans. T. Nugent [1756] (repr., Gainesville, FL: Scholars' Facsimiles and Reprints, 1971).

3. Jean-Jacques Rousseau, *Essay on the Origin of Languages* [1782], trans. John H. Moran and Alexander Gode (Chicago: University of Chicago Press, 1966).

< 217 >

4. Friedrich W. Nietzsche, *Human, All Too Human: A Book for Free Spirits* [1878], trans. R. J. Hollingdale (New York: Cambridge University Press, 1986), 99–100.

5. Wundt actually vies with William James as the founder of experimental psychology, but James was not impressed with his rival. He wrote, "He aims at being a Napoleon of the intellectual world. Unfortunately he will never have a Waterloo, for he is Napoleon without genius and with no central idea." Quoted in Edwin G. Boring, *A History of Experimental Psychology*, 2nd ed. (Englewood Cliffs, NJ: Prentice-Hall, 1950), 113.

6. Wilhelm Wundt, *Die Sprache*, 2 vols. (Leipzig: Enghelman, 1900).

7. McDonald Critchley, *Silent Language* (London: Arnold, 1975), 221.

8. Giorgio Fano, *The Origins and Nature of Language* (Bloomington: Indiana University Press, 1992).

9. Kimon Nikolaïdes, *The Natural Way to Draw* (Boston: Houghton Mifflin, 1975).

10. Gordon W. Hewes, "Primate Communication and the Gestural Origins of Language," *Current Anthropology* 14 (1973: 5–24.

11. Edward S. Klima and Ursula Bellugi, *The Signs of Language* (Cambridge, MA: Harvard University Press, 1979).

12. David F. Armstrong, William C. Stokoe, and Sherman E. Wilcox, *Gesture and the Nature of Language* (Cambridge: Cambridge University Press, 1995).

13. William C. Stokoe, *Language in Hand: Why Sign Came before Speech* (Washington, DC: Gallaudet University Press, 2001).

14. David F. Armstrong, *Original Signs: Gesture, Sign, and the Source of Language* (Washington, DC: Gallaudet University Press, 1999).

15. Michael C. Corballis, *From Hand to Mouth: The Origins of Language* (Princeton, NJ: Princeton University Press, 2002).

16. At a conference on the evolution of language held in Montreal in 2010, the only conclusion on which there was general agreement was that if your first name is Michael, then you believe that language originated in manual gestures. The Michaels in question, who all presented at the conference, were Michael Arbib, Michael Tomasello, and me. Non-Michaels were invited to join, and as we shall see there are many non-Michaels who have seen the light.

17. Justin H. G. Williams et al., "Imitation, Mirror Neurons, and Autism," *Neuroscience and Biobehavioral Reviews* 25 (2001): 287–95.

18. Gregory Hickok, *The Myth of Mirror Neurons* (New York: W. W. Norton, 2014).

19. Giacomo Rizzolatti and Corrado Sinigaglia, "The Functional Role of the Parieto-Frontal Mirror Circuit: Interpretations and Misinterpretations," *Nature Neuroscience* 11 (2010): 264–74.

20. Giacomo Rizzolatti and Michael A. Arbib, "Language within our Grasp," *Trends in Neurosciences* 21 (1998): 188–94.

21. Peter Fonagy and Mary Target, "The Rooting of the Mind in the Body: New Links between Attachment Theory and Psychoanalytic Thought," *Journal of the American Psychoanalytic Association* 55 (2007): 437.

22. Pascal Molenberghs, Ross Cunnington, and Jason B. Mattingley, "Brain Regions with Mirror Properties: A Meta-analysis of 125 Human FMRI Studies," *Neuroscience and Biobehavioral Reviews* 36 (2012): 341–49.

23. Roy Mukamel et al., "Single-Neuron Responses in Humans during Execution and Observation of Actions," *Current Biology* 20 (2010): 750–56.

< 218 >

24. Giacomo Rizzolatti and Corrado Sinigaglia, *Mirrors in the Brain: How Our Minds Share Actions and Emotions* (Oxford: Oxford University Press, 2006).

25. Erica A. Cartmill, Sian Beilock, and Susan Goldin-Meadow, "A Word in the Hand: Action, Gesture, and Mental Representation in Humans and Nonhuman Primates," *Philosophical Transactions of the Royal Society B* 367 (2012): 129–43.

26. Lisa Aziz-Zadeh et al., "Congruent Embodied Representations for Visually Presented Actions and Linguistic Phrases Describing Actions," *Current Biology* 16 (2006): 1818–23.

27. Charles Darwin, *The Descent of Man and Selection in Relation to Sex*, 2nd ed. (New York: Appleton, 1871), 87.

28. Uwe Jürgens, "Neural Pathways Underlying Vocal Control," *Neuroscience and Biobehavioral Reviews* 26 (2002): 235–58.

29. Jane Goodall, *The Chimpanzees of Gombe: Patterns of Behavior* (Cambridge, MA: Harvard University Press, 1986), 125.

30. David Premack, "Human and Animal Cognition: Continuity and Discontinuity," *Proceedings of the National Academy of Sciences (USA)* 104 (2007): 13,866.

31. Adam Clarke Arcadi, Daniel Robert, and Christophe Boesch, "Buttress Drumming by Wild Chimpanzees: Temporal Patterning, Phase Integration into Loud Calls, and Preliminary Evidence for Individual Differences," *Primates* 39 (1998): 505–18.

32. William D. Hopkins, Jared P. Taglialatela, and David A. Leavens, "Chimpanzees Differentially Produce Novel Vocalizations to Capture the Attention of a Human," *Animal Behaviour* 73 (2007): 281–86.

33. Katie E. Slocombe and Klaus Zuberbühler, "Chimpanzees Modify Recruitment Screams as a Function of Audience Composition," *Proceedings of the National Academy of Sciences (USA)* 104 (2007): 17,228–33.

34. Cathy Hayes, *The Ape in Our House* (London, Gollancz, 1952).

35. Sue Savage-Rumbaugh,, Stuart G. Shanker, and Talbot J. Taylor, *Apes, Language, and the Human Mind* (New York: Oxford University Press, 1998).

36. R. Allen Gardner and Beatrice T. Gardner, "Teaching Sign Language to a Chimpanzee," *Science* 165 (1969): 664–72.

37. Serge A. Wich et al., "A Case of Spontaneous Acquisition of a Human Sound by an Orangutan," *Primates* 50 (2009): 56–64.

38. Christopher I. Petkov and Erich D. Jarvis, "Birds, Primates, and Spoken Language Origins: Behavioral Phenotypes and Neurobiological Substrates," *Frontiers in Evolutionary Neuroscience* 4 (2012), article 12, 5.

39. Cristina Colonnesi et al., "The Relation between Pointing and Language Development: A Meta-analysis," *Developmental Review* 30 (2010): 352.

40. Virginia Volterra et al., "Gesture and the Emergence and Development of Language," in *Beyond Nature-Nurture: Essays in Honor of Elizabeth Bates*, ed. Michael Tomasello and Dan Slobin (Mahwah, NJ: Lawrence Erlbaum Associates, 2005), 3–40.

41. Helene Meunier, Jacques Vauclair, and Jacqueline Fagard, "Human Infants and Baboons Show the Same Pattern of Handedness for a Communicative Gesture," *PLoS ONE* 7 (2012), e33559.

42. Michael Tomasello, *The Origins of Human Communication* (Cambridge, MA: MIT Press, 2008).

43. This is challenged by Nicholas J. Mulcahy and Vernon Hedge ("Are Great Apes Tested

with an Abject Object-Choice Task?," *Animal Behaviour* 83 [2012]: 313–21), who review evidence suggesting that chimpanzees, like many other species, can follow human pointing if the objects pointed at are placed relatively far apart.

44. Lee Cronk, "Continuity, Displaced Reference, and Deception," *Behavioral and Brain Sciences* 27 (2004): 510–11.

45. Ulf Liszkowski et al., "Prelinguistic Infants, but Not Chimpanzees, Communicate about Absent Entities," *Psychological Science* 20 (2009): 654–60.

46. Heidi Lyn et al., "Apes Communicate about Absent and Displaced Objects: Methodology Matters," *Animal Cognition* 17 (2014): 85–94.

47. Savage-Rumbaugh, Shanker, and Taylor, *Apes, Language, and the Human Mind*

48. Francine G. P. Patterson and Wendy Gordon, "Twenty-Seven Years of Project Koko and Michael," in *All Apes Great and Small*, vol. 1, *African Apes*, ed. Biruté Galdikas et al. (New York: Kluver, 2001), 165–76.

49. This criticism is itself dubious, because there is no evidence as to whether humans would point if raised in a nonhuman environment.

50. Robin I. M. Dunbar, "Coevolution of Neocortical Size, Group Size, and Language in Humans," *Behavioral and Brain Sciences* 16 (1993): 681–735.

51. Amy S. Pollick and Frans B. M. de Waal, "Apes Gestures and Language Evolution," *Proceedings of the National Academy of Sciences* 104 (2007): 81–89.

52. Cat Hobaiter and Richard W. Byrne, "Serial Gesturing by Wild Chimpanzees: Its Nature and Function for Communication," *Animal Cognition* 14 (2011): 827–38.

53. Cat Hobaiter and Richard W. Byrne, "The Meaning of Chimpanzee Gestures," *Current Biology* 24 (2014): 1–5.

54. Christophe Boesch, "Aspects of Transmission of Tool-Use in Wild Chimpanzees," in *Tools, Language, and Cognition in Human Evolution*, ed. Kathleen R. Gibson and Tim Ingold (Cambridge: Cambridge University Press, 1993), 171–83. Such examples of imitation are anecdotal and are often explainable in other ways.

55. Joanne E. Tanner, Francine G. Patterson, and Richard W. Byrne, "The Development of Spontaneous Gestures in Zoo-Living Gorillas and Sign-Taught Gorillas: From Action and Location to Object Representation," *Journal of Developmental Processes* 1 (2006): 69–103.

56. Anne Russon and Kristin Andrews, "Orangutan Pantomime: Elaborating the Message," *Biology Letters* 7 (2001): 627–30.

57. Simone Pika, Katja Liebal, and Michael Tomasello, "Gestural Communication in Young Gorillas (*Gorilla gorilla*): Gestural Repertoire, and Use," *American Journal of Primatology* 60 (2003): 95–111.

58. Katja Liebal, Josep Call, and Michael Tomasello, "Use of Gesture Sequences in Chimpanzees," *American Journal of Primatology* 64 (2004): 377–96.

59. Simone Pika, Katja Liebal, and Michael Tomasello, "Gestural Communication in Subadult Bonobos (*Pan paniscus*): Repertoire and Use," *American Journal of Primatology* 65 (2005): 39–61.

60. Stephanie L. Bogart and Jill D. Pruetz, "Ecological Context of Savanna Chimpanzee (*Pan troglodyte verus*) Termite Fishing at Fongoli, Senegal," *American Journal of Primatology* 70 (2008): 605–12.

61. Jill D. Pruetz and Paco Bertolani, "Savanna Chimpanzees, *Pan troglodytes verus*, Hunt with Tools," *Current Biology* 17 (2007): 412–17.

< 220 >

62. Christophe Boesch, Josephine Head, and Martha M. Robbins, "Complex Tool Sets for Honey Extraction among Chimpanzees in Laongo National Park, Gabon," *Journal of Human Evolution* 56 (2009): 560–69.

63. Tomasello, *Origins of Human Communication*, 55.

64. Esteban Rivas, "Recent Use of Signs by Chimpanzees (*Pan troglodytes*) in Interactions with Humans," *Journal of Comparative Psychology* 119 (2005): 404–17.

65. Patricia Greenfield and E. Sue Savage-Rumbaugh, "Grammatical Combination in *Pan paniscus*: Processes of Learning and Invention in the Evolution and Development of Language," in *"Language" and Intelligence in Monkeys and Apes*, ed. Sue Taylor Parker and Kathleen Rita Gibson (Cambridge: Cambridge University Press, 1990), 540–78.

66. David A. Leavens and Timothy P. Racine, "Joint Attention in Apes and Humans," *Journal of Consciousness Studies* 16 (2000): 240–67.

67. Elizabeth Pennisi, "The Burdens of Being a Biped," *Science* 336 (2012): 974.

68. Dennis M. Bramble and Daniel E. Lieberman, "Endurance Running and the Evolution of *Homo*," *Science* 432 (2004): 345–52.

69. Mark Verhaegen, "The Aquatic Ape Evolves: Common Misconceptions and Unproven Assumptions about the So-Called Aquatic Ape Hypothesis," *Human Evolution* 28 (2013): 237–66.

70. C. Owen Lovejoy et al., "The Pelvis and Femur of *Ardipithecus Ramidus*: The Emergence of Upright Walking," *Science* 326 (2009): 71, 71e1–71e6. The bones of the foot of another hominin, dated at 3.4 million years ago, reveals an opposable toe, suggesting part adaptation for climbing as well as for bipedal walking. Those feet in ancient times were more multipurpose than they are today.

71. Suzanna K. S. Thorpe, Roger L. Holder, and Robin H. Crompton, "Origin of Human Bipedalism as an Adaptation for Locomotion on Flexible Branches," *Science* 316 (2007): 1328–31.

72. Glen McBride, "Story Telling, Behavior Planning, and Language Evolution in Context." *Frontiers in Psychology* 5 (2014), article 1131, 3.

73. Jiang Xu et al., "Symbolic Gestures and Spoken Language Are Processed by a Common Neural System," *Proceedings of the National Academy of Sciences* 106 (2009): 20,664–69.

74. Paola Pietrandrea, "Iconicity and Arbitrariness in Italian Sign Language," *Sign Language Studies* 2 (2002): 296–321.

75. Karen Emmorey, *Language, Cognition, and Brain: Insights from Sign Language Research* (Hillsdale, NJ: Erlbaum, 2002).

76. Jane Marshall et al., "Aphasia in a User of British Sign Language: Dissociation between Sign and Gesture," *Cognitive Neuropsychology* 21 (2004): 537–54.

77. Elena Pizzuto and Virginia Volterra, "Iconicity and Transparency in Sign Languages: A Cross-Linguistic Cross-Cultural View," in *The Signs of Language Revisited: An Anthology to Honor Ursula Bellugi and Edward Klima*, ed. Karen Emmorey and Harlan Lane (Mahwah, NJ: Lawrence Erlbaum Associates, 2000), 261–86.

78. Nancy Frishberg, "Arbitrariness and Iconicity in American Sign Language," *Language* 51 (1975): 696–719.

79. Charles Darwin, *The Expression of the Emotions in Man and Animals* (London: John Murray, 1872), 62. I'm not quite sure why he included "savages."

80. Ann Senghas, Sotaro Kita, and Asli Özyürek, "Children Creating Core Properties of

< 221 >

Language: Evidence from an Emerging Sign Language in Nicaragua," *Science* 305 (2004): 1780–82.

81. Ulrike Zeshan, "Sign Language in Turkey: The Story of a Hidden Language," *Turkic Languages* 6 (2002): 229–74.

82. Nicholas Evans, *Dying Words: Endangered Languages and What They Have to Tell Us* (Oxford: Wiley-Blackwell, 2009).

83. Mark Aronoff et al., "The Roots of Linguistic Organization in a New Language," *Interaction Studies* 9 (2008): 133–53.

84. Tomasello, *Origins of Human Communication*, 246.

CHAPTER EIGHT

1. Lewis Carroll, *Sylvie and Bruno* (London: Macmillan), 83.

2. Robbins Burling, *The Talking Ape* (New York: Oxford University Press, 2005), 123.

3. To be more precise, the circuitry for both vocalization and pectoral appendages originates in the same rhombomere 8-spinal compartment. The coupling between vocal and pectoral sound-producing systems can be traced through fish, birds, frogs, and primates (Andrew H. Bass and Boris P. Chagnaud, "Shared Developmental and Evolutionary Origins for Neural Basis of Vocal-Acoustic and Pectoral-Gestural Signaling," *Proceedings of the National Academy of Sciences (USA)* 109 [2012]: 10,677–84).

4. David McNeill, "So You Think Gestures Are Nonverbal?," *Psychological Review* 92 (1985): 370.

5. Alvin Liberman et al., "Perception of the Speech Code," *Psychological Review* 92 (1985): 431–61.

6. So I now do all my own typing, with just two fast-moving fingers.

7. Technically, letters are known as graphemes, and the units of sounds as phonemes. But phonemes defy simple analysis in terms of sound, even though we seem to hear them distinctly enough.

8. Geoffrey Hinton et al, "Deep Neural Networks for Acoustic Modeling in Speech Recognition," *IEEE Signal Processing Magazine* 29 (2012): 82–97.

9. Evelyne Kohler et al., "Hearing Sounds, Understanding Actions: Action Representation in Mirror Neurons," *Science* 297 (2002): 846–88. These results also suggest that mirror-neuron activity is learned rather than inborn. Monkeys are probably not innately endowed with the ability to recognize paper tearing. Kittens are another story.

10. Jan Fritz, Mortimer Mishkin, and Richard C. Saunders, "In Search of an Auditory Engram," *Proceedings of the National Academy of Sciences USA* 102 (2005): 9359–64. Of course monkeys do make a wide variety of calls, but these are largely instinctive and don't rely on learning and memory.

11. Katrin Schulze, Faraneh Vargha-Khadem, and Mortimer Mishkin, "Test of a Motor Theory of Long-Term Auditory Memory," *Proceedings of the National Academy of Sciences* 109 (2012): 7121. You may wonder about our ability to remember music, though. I suspect musical memory is much better if we play an instrument or if words are associated with the tune. Even humming a tune may help consolidate memory of it.

12. Silvia Benavides-Varela et al., "Newborn's Brain Activity Signals the Origin of Word Memories," *Proceedings of the National Academy of Sciences (USA)* 109 (2012): 17,908–13.

13. The motor theory has long been controversial, with evidence both for it and in favor of

< 222 >

the alternative view that the perception of speech is purely auditory. For a review of evidence and the compromise suggested here, see Vittorio Gallese et al., "Mirror Neuron Forum," *Perspectives on Psychological Science* 6 (2011): 369–407.

14. Gregory Hickok and David Poeppel, "The Cortical Organization of Speech Processing," *Nature Reviews Neuroscience* 8 (2007): 393–402.

15. Mortimer Mishkin, Leslie G. Ungerleider, and Kathleen A. Macko, "Object Vision and Spatial Vision: Two Cortical Pathways," *Trends in Neuroscience* 6 (1983): 414–7.

16. A. David Milner and Melvyn A. Goodale, *The Visual Brain in Action*, 2nd ed. (Oxford: Oxford University Press, 2006).

17. Andreas R. Pfenning et al., "Convergent Specializations in the Brains of Humans and Song-Learning Birds," *Science* 346 (2014): 1333–46.

18. Michael Petrides and Deepak N. Pandya, "Distinct Parietal and Temporal Pathways to the Homologue of Broca's Area in the Monkey," *PLoS ONE* 7(8) (2009), e1000170, 11.

19. Gregory Hickok, "Eight Problems for the Mirror Neuron Theory of Action Understanding in Monkeys and Humans," *Journal of Cognitive Neuroscience* 21 (2009): 1229–43.

20. Ferdinand de Saussure, *Course in General Linguistics* [*Cours de linguistique générale*, 1916], translated by W. Baskin (Glasgow: Fontana/Collins, 1977).

21. Wolfgang Köhler, *Gestalt Psychology* (New York: Liveright, 1929).

22. Vilayanur S. Ramachandran and Edward M. Hubbard, "Synaesthesia—A Window into Perception, Thought and Language," *Journal of Consciousness Studies* 8 (2001): 3–34.

23. Richard A. S. Paget, "The Origin of Speech—A Hypothesis," *Proceedings of the Royal Society of London A*, 119 (1928): 157–72.

24. Matthias Urban, "Conventional Sound Symbolism in Terms for Organs of Speech: A Cross-Linguistic Study," *Folia Linguistica* 45 (2011): 199–214.

25. Hadas Shintel, Howard C. Nusbaum, and Arika Okrent, "Analog Acoustic in Speech," *Journal of Memory and Language*, 55 (2006): 167–77.

26. Hadas Shintel and Howard C. Nusbaum, "The Sound of Motion in Spoken Language: Visual Information Conveyed by Acoustic Properties of Speech," *Cognition* 105 (2007): 681–90.

27. Sarah Dolscheid et al., "Prelinguistic Infants Are Sensitive to Space-Pitch Associations Found across Cultures," *Psychological Science* 25 (2014): 1256–61.

28. Steven Pinker, *The Stuff of Thought* (London: Penguin Books, 2007).

29. Ramachandran and Hubbard, "Synaesthesia," 19.

30. Paget, "Origin of Speech," 170.

31. Pamela Perniss and Gabriella Vigliocco, "The Bridge of Iconicity: From a World of Experience to the Experience of Language," *Philosophical Transactions of the Royal Society B* 369 (2014): 20130300.

32. Now a rather shallow river in Italy, but more of a challenge in Julius Caesar's day.

33. Seth Dobson, "Allometry of Facial Mobility in Anthropoid Primates: Implications for the Evolution of Facial Expression," *American Journal of Physical Anthropology* 138 (2009): 70–81.

34. To whom she said it is not recorded, as far as I know, but it may have been to the bishop.

35. Seth D. Dobson and Chet C. Sherwood, "Correlated Evolution of Brain Regions Involved in Producing and Processing Facial Expressions in Anthropoid Primates," *Biology Letters* 7 (2011): 86–88.

< 223 >

36. Rachel Sutton-Spence and Penny Boyes-Braem, eds., *The Hands Are the Head of the Mouth: The Mouth as Articulator in Sign Language* (Hamburg: Signum-Verlag, 2001).

37. Laura J. Muir and Iain E. G. Richardson, "Perception of Sign Language and Its Application to Visual Communication for Deaf People," *Journal of Deaf Studies and Deaf Education* 10 (2005): 390–401.

38. In modern times, the haka is perhaps best known internationally as a ritual performed by the New Zealand rugby team, the All Blacks, before every international match. Their status as world champions, it is sometimes suggested, may depend on their having intimidated their opponents before the match is played.

39. Harry McGurk and John MacDonald, "Hearing Lips and Seeing Voices," *Nature* 264 (1976): 746–48. At the time of writing (July 2014), and I hope still, you could experience the McGurk effect on www.youtube.com/watch?v=aFPtc8BVdJk.

40. Gemma A. Calvert and Ruth Campbell, "Reading Speech from Still and Moving Faces: The Neural Substrates of Visible Speech," *Journal of Cognitive Neuroscience* 15:57–70.

41. Robin Dunbar, "Bridging the Bonding Gap: The Transition from Primates to Humans," *Proceedings of the Royal Society B* 367 (2012): 1837–46.

42. Brian Villmoare et al., "Early Homo at 2.8 Ma from Ledi-Geraru, Afar, Ethiopia," *Science Express*, March 5, 2015, doi 10.1126/science.aaa1343.

43. Giacomo Rizzolatti et al., "Functional Organization of Inferior Area 6 in the Macaque Monkey, Part 2, Area F5 and the Control of Distal Movements," *Experimental Brain Research* 71 (1988): 491–507.

44. Maurizio Gentilucci et al., "Grasp with Hand and Mouth: A Kinematic Study on Healthy Subjects," *Journal of Neurophysiology* 86 (2001): 1685–99. You are advised to try to keep your mouth shut when reaching for a large object, such as a pineapple. Or even when trying to grasp a big idea, which can cause the mouth to drop.

45. Maurizio Gentilucci and Michael C. Corballis, "From Manual Gesture to Speech: A Gradual Transition," *Neuroscience and Biobehavioral Reviews* 30 (2006): 949–60.

46. Paolo Bernardis et al., "Manual Actions Affect Vocalizations of Infants," *Experimental Brain Research* 184 (2008): 599–603.

47. Bailey A. Russell, Frank J. Cerny, and Elaine T. Stathopoulos, "Effects of Varied Vocal Intensity on Ventilation and Energy Expenditure in Women and Men," *Journal of Speech, Language, and Hearing Research* 41 (1998): 239–48.

48. Charles Darwin, *The Descent of Man and Selection in Relation to Sex*, 2nd ed. (New York: Appleton, 1896), 78.

49. Mary Kingsley, *Travels in West Africa, Cong Française, Corisco, and Cameroons* (London: F. Cass, 1897), 504.

50. Peter F. MacNeilage, *The Origin of Speech* (Oxford: Oxford University Press, 2008).

51. An individual can be said to have a 1% advantage in fitness over another individual if he or she produces an average of 1.01 offspring for every 1.00 offspring produced by the other. John B. S. Haldane, "A Mathematical Theory of Natural and Artificial Selection, Part 5, Selection and Mutation," *Proceedings of the Cambridge Philosophical Society* 23 (1927): 838–44.

52. Ezra B. Zubrow, "The Demographic Modeling of Neanderthal Extinction," in *The Human Revolution: Behavioral and Biological Perspectives on the Origins of Modern Humans*, ed. Paul Mellars and Chris Stringer (Princeton NJ: Princeton University Press, 1989), 212–31.

53. Adam Kendon, "Vocalisation, Speech, Gesture, and the Language Origins Debate," *Gesture* 13 (2011): 349–70. I have responded to Kendon in Michael C. Corballis, "The Word Ac-

< 224 >

cording to Adam: The Role of Gesture in Language Evolution," in *From Gesture in Conversation to Visible Action as Utterance: Essays in Honor of Adam Kendon*, ed. Mandana Seyfeddinipur and Marianne Gullberg (Amsterdam: John Benjamins, 2014), 177–97.

54. David McNeill, *How Language Began: Gesture and Speech in Human Evolution* (Cambridge: Cambridge University Press, 2012).

55. Charles F. Hockett, "In Search of Love's Brow," *American Speech* 53 (1978): 274–75.

56. Mark Verhaegen, "The Aquatic Ape Evolves: Common Misconceptions and Unproven Assumptions about the So-Called Aquatic Ape Hypothesis," *Human Evolution* 28 (2013): 238.

57. Hervé Bocherens, Gennady Baryshnikov, and Wim Van Neer, "Were Bears or Lions Involved in Salmon Accumulation in the Middle Palaeolithic of the Caucasus? An Isotopic Investigation in Kudaro 3 Cave," *Quaternary International* 339–40 (2014): 112–18.

58. Philip Lieberman, "The Evolution of Human Speech," *Current Anthropology* 48 (2007): 39–46.

59. At the time of writing, this was available for your listening pleasure on www.fau.edu/explore/homepage-stories/2008-04speaks.php.

60. Daniel E. Lieberman, *The Story of the Human Body: Evolution, Health and Disease* (New York: Pantheon, 2013).

61. Daniel E. Lieberman, "Sphenoid Shortening and the Evolution of Modern Cranial Shape," *Nature* 393 (1998): 158–62.

62. Daniel E. Lieberman, Brandeis M. McBratney, and Gail Krovitz, "The Evolution and Development of Cranial Form in Homo Sapiens," *Proceedings of the National Academy of Sciences* 99 (2002): 1134–39.

63. P. Lieberman, "Evolution of Human Speech," 39.

64. Jean-Louis Boë et al., "The Vocal Tract of Newborn Humans and Neanderthals: Acoustic Capabilities and Consequences for the Debate on the Origin of Language; A Reply to Lieberman (2007a)," *Journal of Phonetics* 35 (2007): 564–81.

65. Bart De Boer and W. Tecumseh Fitch, "Computer Models of Vocal Tract Evolution: An Overview and Critique," *Adaptive Behaviour* 18 (2010): 36–47.

66. Anna Barney et al., "Articulatory Capacity of Neanderthals: A Very Recent and Human-Like Fossil Hominin," *Philosophical Transactions of the Royal Society B* 367 (2012): 88–102.

67. Faraneh Vargha-Khadem et al., "Praxic and Nonverbal Cognitive Deficits in a Large Family with a Genetically Transmitted Speech and Language Disorder," *Proceedings of the National Academy of Sciences, USA* 92 (1995): 930–33.

68. Simon E. Fisher, "Localisation of a Gene Implicated in a Severe Speech and Language Disorder," *Nature Genetics* 18 (1998): 168–70.

69. Myrna Gopnik, "Feature-Blind Grammar and Dysphasia," *Nature* 344 (1990): 715; Steven Pinker, *The Language Instinct* (New York: William Morrow, 1994).

70. Frédérique Liégeois et al., "Language fMRI Abnormalities Associated with FOXP2 Gene Mutation," *Nature Neuroscience* 6 (2003): 1230–37.

71. Graham Coop et al., "The Timing of Selection of the Human FOXP2 gene," *Molecular Biology and Evolution* 25 (2008): 1257–59.

72. Johannes Krause et al., "The Derived FOXP2 Variant of Modern Humans Was Shared with Neandertals," *Current Biology* 17 (2007): 1908–12.

73. Susan E. Ptak et al., "Linkage Disequilibrium Extends across Putative Selected Sites in FOXP2," *Molecular and Biological Evolution* 26 (2012): 2181–84.

< 225 >

74. Sebastian Haesler et al., "Incomplete and Inaccurate Vocal Imitation after Knockdown of FOXP2 in Songbird Basal Ganglia Nucleus Area X," *PLoS Biology* 5 (2007): 2885–97.

75. Wolfgang Enard et al., "A Humanized Version of Foxp2 Affects Cortico-Basal Ganglia Circuits in Mice," *Cell* 137 (2009): 968–69. But perhaps they should not be too confident; for more recent evidence see Kurt Hammerschmidt et al., "A Humanized Version of Foxp2 Does Not Affect Ultrasonic Vocalization in Adult Mice." *Genes, Brain, and Behavior* 14 (2015), pre-publication download at doi 10.1111/gbb.12237.

76. Weiguo Shu et al., "Characterization of a New Subfamily of Winged-Helix/Forkhead (Fox) Genes that are Expressed in the Lung and Act as Transcriptional Repressors," *Journal of Biological Chemistry* 276 (2001): 27,488–97.

77. John E. Pfeiffer, *The Emergence of Man* (London: Book Club Associates, 1973), 160.

78. Paolo Villa and Wil Roebroeks, "Neandertal Demise: An Archaeological Analysis of the Modern Human Superiority Complex," *PLoS ONE* 9 (2014), e96424, 1.

79. Sverker Johansson, "The Talking Neandertals: What Do Fossils, Genetics, and Archeology Say?" *Biolinguistics* 7 (2013): 35–74.

80. Dan Dediu and Stephen C. Levinson, "On the Antiquity of Language: The Reinterpretation of Neandertal Linguistic Capacities and its Consequences," *Frontiers in Psychology* 4 (2013), article 397.

81. Joaquin Rodríguez-Vidal et al., "A Rock Engraving Made by Neanderthals in Gibraltar," *Proceedings of the National Academy of Sciences* 111 (2014): 13,301–6.

82. Do not be too hasty, though—see Thomas Wynn, Karenleigh A. Overmann, and Frederick L. Coolidge, "The False Dichotomy: A Refutation of the Neandertal Indistinguishability Claim," *Journal of Anthropological Sciences* 94 (2016): 1–22. This is a spirited defense of the fundamental differences between us and the Neandertals.

CHAPTER NINE

1. Benjamin Lee Whorf, "Science and Linguistics," *Technology Review* 42 (1940): 227–31, 247–48.

2. Words can be broken down in to small units of meaning called *morphemes*; thus a word like *preordain* can be seen as made up of the morphemes *pre* and *ordain*. But from developmental and perhaps evolutionary perspectives, the word itself is probably primary—see Mark Aronoff, "In the Beginning Was the Word," *Language* 83 (2007): 803–30. Kids learn words before they understand (if they ever do) how they can be broken down into morphemes. James Hurford (*The Origins of Grammar: Language in the Light of Evolution II* [Oxford: Oxford University Press, 2012]) suggests that the earliest languages were made up of words without internal structure.

3. Marcos M. Duarte, "Red Ochre and Shells: Clues to Human Evolution," *Trends in Ecology and Evolution* 29 (2014): 560–65.

4. Cited by in ibid., 562.

5. Paul Bloom, *How Children Learn the Meanings of Words* (Cambridge, MA: MIT Press, 2000)—give or take 10,000 or so, depending on whether you are a Pinker or a Hurford. Or perhaps a psychologist or a linguist.

6. H. Lyn White Miles and Stephen E. Harper, "Ape Language Studies and the Study of Human Language Origins," in *Hominid Culture in Primate Perspective*, ed. Duane D. Quiatt and Junichiro Itani (Niwot: University Press of Colorado, 1994), 253–78.

< 226 >

7. Juliane Kaminski, Josep Call, and Julia Fischer, "Word Learning in the Domestic Dog: Evidence for 'Fast Mapping,'" *Science* 304 (2004): 1682–83. For a critical comment, though, see Ulrike Giebel and D. Kimbrough Oller, "Vocabulary Learning in a Yorkshire Terrier: Slow Mapping of Spoken Words." *PLoS ONE* 7 (2012):, article30182; these authors describe the exploits of another dog, this time a Yorkshire terrier called Betty. She has a receptive vocabulary of 117 words, but the authors prefer to describe her acquisition, and that of Rico and Chaser, as "slow mapping."

8. Some linguists have complained to me that these examples of dogs and apes apparently understanding words does not mean that they understand words as symbols. It's true that these animal can't produce words, and there is no evidence that they can use the words to construct sentences or to reason. All I am concerned with here, though, is the association with objects or actions in the real world. The combining of words to form or understand sentences takes us to another level of language. In an evolutionary account, you can't really expect everything to be present at once—although in Chomsky's account everything *does* seem to happen at once!

9. Robert M. Seyfarth, Dorothy L. Cheney, and Peter Marler, "Monkey Responses to Three Different Alarm Calls—Evidence of Predator Classification and Semantic Communication," *Science* 210 (1980): 801–3.

10. Dorothy L. Cheney and Robert M. Seyfarth, *How Monkeys See the World* (Chicago: University of Chicago Press, 1990).

11. From the 1894 poem "Two Loves," by Lord Alfred Douglas, mentioned at the trial of Oscar Wilde for "gross indecency." We've moved on from there.

12. Except in Australia, where it's a beach.

13. Victoria C. Martin et al., "A Role for the Hippocampus in Encoding Simulations of Future Events," *Proceedings of the National Academy of Sciences* 108 (2011): 13,858–63.

14. Milan Kundera, *Ignorance*, trans. L. Asher (New York: HarperCollins, 2002), 122–23.

15. These were *To the Is-Land* (1982), *An Angel at My Table* (1984), and *The Envoy from Mirror City* (1985). They were collected in a single volume as Janet Frame, *An Autobiography* (Auckland: Century Hutchinson, 1989). They also formed the basis of the 1990 movie *An Angel at My Table*, directed by Jane Campion.

16. Steven Pinker, "Language as an Adaptation to the Cognitive Niche," in *Language Evolution*, ed. Morten H. Christiansen and Simon Kirby (Oxford: Oxford University Press, 2003), 16–37.

17. These are the names of my granddaughters, who actually get along pretty well.

18. Ann Senghas, Sotaro Kita, and Asli Özyürek, "Children Creating Core Properties of Language: Evidence from an Emerging Sign Language in Nicaragua," *Science* 305 (2004): 1780–82.

19. Simon Kirby and James R. Hurford, "The Emergence of Linguistic Structure: An Overview of the Iterated Learning Model," in *Simulating the Evolution of Language*, ed. Angelo Cangelosi and Domenico Parisi (London: Springer Verlag, 2002), 121–48.

20. Murray Gell-Mann and Merritt Ruhlen, "The Origin and Evolution of Word Order," *Proceedings of the National Academy of Sciences USA* 108 (2011): 17,290–95. These authors evidently assume that the earliest language was spoken and do not consider the possibility that earlier languages were signed.

21. Wolcott Gibbs, "Time . . . Fortune . . . Life . . . Luce," *New Yorker*, November 28, 1936, 20–25.

< 227 >

22. Paul J. Hopper and Elizabeth Closs Traugott, *Grammaticalization*, 2nd ed. (Cambridge: Cambridge University Press, 2003).

23. Or so I have been informed, possibly unreliably.

24. Paul D. Deane, *Grammar in Mind and Brain: Explorations in Cognitive Syntax* (Berlin: Mouton de Gruyter, 1992).

25. Paul D. Deane, "Neurological Evidence for a Cognitive Theory of Syntax: Agrammatic Aphasia and the Specialization of Form Hypothesis," in *Cognitive Linguistics in the Redwoods: The Expansion of a New Paradigm in Linguistics*, ed. Eugene H. Casad (Berlin: Mouton de Gruyter, 1995), 55–115.

26. John O'Keefe, "The Spatial Prepositions in English, Vector Grammar, and the Cognitive Map Theory," in *Language and Space*, ed. Paul Bloom et al. (Cambridge, MA: MIT Press, 1996), 277–316.

27. Virginia Frisk and Brenda Milner, "The Role of the Left Hippocampal Region in the Acquisition and Retention of Story Content," *Neuropsychologia* 28 (1990): 349–59.

28. Melissa C. Duff and Sarah Brown-Schmidt, "The Hippocampus and the Flexible Use and Processing of Language," *Frontiers in Human Neuroscience* 6 (2012), article 69, 1.

29. Deane, "Neurological Evidence," 62.

30. Dietrich Stout and Thierry Chaminade, "Stone Tools, Language and the Brain in Human Evolution," *Philosophical Transactions of the Royal Society B: Biological Sciences* 367 (2012): 75–87.

31. Jane Wilkins et al., "Evidence for Early Hafted Hunting Technology," *Science* 338 (2012): 942–46.

32. Ian Tattersall, *Masters of the Planet: The Search for Human Origins* (New York: Palgrave Macmillan, 2012).

33. Katarina Pastra and Yiannis Aloiminos ("The Minimalist Grammar of Action," *Philosophical Transactions of the Royal Society B* 367 [2012]: 103–17) go further and develop a theory of tool use based on Chomsky's Minimalist Program and Merge operation. They don't seem to claim that tool use was prior to language but suggest rather that they share the same structures.

34. Stout and Chaminade, "Stone Tools," 82.

35. Daniel Dor, *The Instruction of Imagination: Language as a Social Communication Technology* (New York: Oxford University Press, 2015).

36. Daniel L. Everett, *Language: The Cultural Tool* (New York: Pantheon, 2012).

37. Alistair Knott, *Sensorimotor Cognition and Natural Language Syntax* (Cambridge, MA: MIT Press, 2012), 624.

CHAPTER TEN

1. Plutarch, "Life of Pompeius," in *Plutarch's Lives*, ed. Aubrey Stewart and George Long, trans. George Long (London: George Bell and Sons, 1892), 270.

2. Jean Poole and Denis G. Lander, "The Pigeon's Concept of Pigeon," *Psychonomic Science* 25 (1971): 157–58.

3. In a recent cricket match between New Zealand and Australia, one of the Australian players complained that his team was put off their game because the New Zealanders were too nice. We all have our crosses to bear.

4. Well, that's according to Richard Dawkins, *The Selfish Gene*, 2nd ed. (Oxford: Oxford University Press, 1989), who named the twelfth chapter of his book "Nice Guys Finish First."

< 228 >

The more common version is "Nice guys finish last," although some versions have it as "Nice guys come second." Take your pick, but be nice about it.

5. Hemlata Pande and Hoshiya S. Dhami, "Model Generation for Word Length Frequencies in Texts with the Application of Zipf's Order Approach," *Journal of Quantitative Linguistics* 19 (2012): 249–61.

6. Ramon Ferrer i Cancho and Ricard V. Solé, "Least Effort and the Origins of Scaling in Human Language," *Proceedings of the National Academy of Sciences (USA)* 100 (2003): 788–91. Although these authors apply the principle of least effort to speech, it can probably be applied to the whole sweep of language evolution, from mime to speech, and even within signed languages themselves. Why are three-letter words the most common? Single-letter and two-letter words are limited by the number of possibilities available, and it seems that it's only when you get to three letters that you can generate enough different combinations for good coverage of what's important—*Mom*, *Dad*, *dog*, *cat*, *sis*, and (in New Zealand) *bro*. After that it's all downhill.

7. Michael C. Corballis, "The Origins of Modernity: Was Autonomous Speech the Critical Factor?," *Psychological Review* 111 (2004): 543–52.

8. Eörs Szathmàry, "Toward Major Evolutionary Transitions 2.0," *Proceedings of the National Academy of Sciences* 112 (2015): 10,109.

9. John F. Hoffecker, "Innovation and Technological Knowledge in the Upper Paleolithic," *Evolutionary Anthropology* 14 (2005: 186–98.

10. Brenna M. Henn, Luigi Luca Cavalli-Sforza, and Marcus W. Feldman, "The Great Human Expansion," *Proceedings of the National Academy of Sciences USA* 109 (2012): 17,758–64.

11. Eva Schultze-Berndt, "Simple and Complex Verbs in Jaminjung: A Study of Event Categorization in an Australian Language," PhD diss., Radboud University, MPI Series in Psycholinguistics, 2000.

12. Daniel L. Everett, *Language: The Cultural Tool* (New York: Pantheon, 2012).

13. Actually, I should say children's children, since it's my grandchildren who are of the ages to learn these things.

< 229 >

Bibliography

Addis, Donna Rose, Alana T. Wong, and Daniel L. Schachter. "Remembering the Past and Imagining the Future: Common and Distinct Neural Substrates during Event Construction and Elaboration." *Neuropsychologia* 45 (2007): 1363–77.

Alexander, Richard D. *How Did Humans Evolve? Reflections on the Uniquely Unique Species*. Museum of Zoology Special Publication 1. Ann Arbor: University of Michigan, 1990.

Almea, Charlotte B., et al. "Place Cells in the Hippocampus: Eleven Maps for Eleven Rooms." *Proceedings of the National Academy of Sciences (USA)* 111 (2014): 18,428–35.

Arbib, Michael A. *How the Brain Got Language: The Mirror System Hypothesis*. Oxford: Oxford University Press, 2012.

Arcadi, Adam Clark, Daniel Robert, and Christophe Boesch. "Buttress Drumming by Wild Chimpanzees: Temporal Patterning, Phase Integration into Loud Calls, and Preliminary Evidence for Individual Differences." *Primates* 39 (1998): 505–18.

Aristotle. *Poetics*. Ann Arbor: University of Michigan Press, 1970.

Armstrong, Davis F. *Original Signs: Gesture, Sign, and the Source of Language*. Washington, DC: Gallaudet University Press, 1999.

Armstrong, David F., William C. Stokoe, and Sherman E. Wilcox. *Gesture and the Nature of Language*. Cambridge: Cambridge University Press, 1995.

Aronoff, Mark. "In the Beginning Was the Word." *Language* 83 (2007): 803–30.

Aronoff, Mark, Irit Meir, Carol A. Padden, and Wendy Sandler. "The Roots of Linguistic Organization in a New Language." *Interaction Studies* 9 (2008): 133–53.

Arsuaga, Juan Luis, et al. "Neandertal Roots: Cranial and Chronological Evidence from Sima de los Huesos." *Science* 344 (2014): 1356–73.

Atkinson, Quentin D. "Phonemic Diversity Supports a Serial Founder Effect Model of Language Expansion from Africa." *Science* 332 (2011): 346–49.

Auel, Jean M. *Clan of the Cave Bear*. London: Hachette UK, 1980.

Aziz-Zadeh, Lisa, Stephen M. Wilson, Giacomo Rizzolatti, and Marco Iacoboni. "Congruent Embodied Representations for Visually Presented Actions and Linguistic Phrases Describing Actions." *Current Biology* 16 (2006): 1818–23.

Barney, Anna, Sandra Martelli, Antoine Serrurier, and James Steele. "Articulatory Capacity of Neanderthals: A Very Recent and Human-Like Fossil Hominin." *Philosophical Transactions of the Royal Society B* 367 (2012): 88–102.

Bass, Andrew H., and Boris P. Chagnaud. "Shared Developmental and Evolutionary Origins

< 231 >

for Neural Basis of Vocal-Acoustic and Pectoral-Gestural Signaling." *Proceedings of the National Academy of Sciences (USA)* 109 (2012): 10,677–84.

Behme, Christina. "A 'Galilean' Science of Language." *Journal of Linguistics* 50 (2014): 671–704.

Benavides-Varela, Silvia, et al. "Newborn's Brain Activity Signals the Origin of Word Memories." *Proceedings of the National Academy of Sciences (USA)* 109 (2012): 17,908–13.

Bernardis, Paolo, et al. "Manual Actions Affect Vocalizations of Infants." *Experimental Brain Research* 184 (2008): 599–603.

Bickerton, Derek. *More than Nature Needs: Language, Mind, and Evolution.* Cambridge, MA: Harvard University Press, 2014.

Bloom, Paul. *How Children Learn the Meanings of Words.* Cambridge, MA: MIT Press, 2000.

Bocherens, Hervé, Gennady Baryshnikov, and Wim Van Neer. "Were Bears or Lions Involved in Salmon Accumulation in the Middle Palaeolithic of the Caucasus? An Isotopic Investigation in Kudaro 3 Cave." *Quaternary International* 339–40 (2014): 112–18.

Bod, Rens. *Beyond Grammar: An Experience-Based Theory of Grammar.* Stanford, CA: CSLI [Center for the Study of Language and Information] Publications, 1998.

———. "From Exemplar to Grammar: A Probabilistic Analogy–Based Model of Language Learning." *Cognitive Science* 33 (2009): 752–93.

Boë, Louis-Jean, et al. "The Vocal Tract of Newborn Humans and Neanderthals: Acoustic Capabilities and Consequences for the Debate on the Origin of Language; A Reply to Lieberman (2007a)." *Journal of Phonetics* 35 (2007): 564–81.

Boesch, Christophe. "Aspects of Transmission of Tool-Use in Wild Chimpanzees." In *Tools, Language, and Cognition in Human Evolution*, edited by Kathleen R. Gibson and Tim Ingold, 171–83. Cambridge: Cambridge University Press, 1993.

Boesch, Christophe, Josephine Head, and Martha M. Robbins. "Complex Tool Sets for Honey Extraction among Chimpanzees in Laongo National Park, Gabon." *Journal of Human Evolution* 56 (2009): 560–69.

Bogart, Stephanie L., and Jill D. Pruetz. "Ecological Context of Savanna Chimpanzee (*Pan troglodyte verus*) Termite Fishing at Fongoli, Senegal." *American Journal of Primatology* 70 (2008): 605–12.

Borges, Jorge Luis. *The Library of Babel.* Translated by J.E.I [James Irby] from *El jardín de senderos que se bifurcan* (1941). http://maskofreason.files.wordpress.com/2011/02/the-library-of-babel-by-jorge-luis-borges.pdf.

Boring, Edwin G. *A History of Experimental Psychology.* 2nd ed. Englewood Cliffs, NJ: Prentice-Hall, 1950.

Boyd, Brian. *The Origin of Stories: Evolution, Cognition, and Fiction.* Cambridge, MA: Belknap Press of Harvard University Press, 2009.

Bramble, Dennis M., and Daniel E. Lieberman. "Endurance Running and the Evolution of *Homo*." *Science* 432 (2004): 345–52.

Bruner, Jerome. *Actual Minds, Possible Worlds.* Cambridge, MA: Harvard University Press, 1986.

Buckner, Randy L., Jessica R. Andrews-Hanna, and Daniel L. Schacter. "The Brain's Default Network: Anatomy, Function, and Relevance to Disease." *Annals of the New York Academy of Sciences* 1124 (2008): 1–38.

Burkart, Judith Maria, and Adolf Heschl. "Perspective Taking or Behaviour Reading? Under-

< 232 >

standing of Visual Access in Common Marmosets (*Calithrix jacchus*)." *Animal Behaviour* 73 (2007): 457–69.

Burling, Robbins. "Motivation, Conventionalization, and Arbitrariness in the Origin of Language." In *The Origins of Language: What Human Primates Can Tell Us*, edited by Barbara J. King, 307–50. Santa Fe, NM: School of American Research Press, 1999.

———. *The Talking Ape*. New York: Oxford University Press, 2005.

Cain, Susan. *Quiet: The Power of Introverts in a World That Can't Stop Talking*. London: Penguin Books, 2013.

Call, Josep, and Michael Tomasello. "Does the Chimpanzee Have a Theory of Mind? 30 Years Later." *Trends in Cognitive Science* 12 (2008): 187–92.

Calvert, Gemma A., and Ruth Campbell. "Reading Speech from Still and Moving Faces: The Neural Substrates of Visible Speech." *Journal of Cognitive Neuroscience* 15:57–70.

Carreiras, Manuel, Jorge Lopez, Francisco Rivero, and David Corina. "Neural Processing of a Whistled Language." *Nature* 433 (2005): 31–32.

Carroll, Lewis. *Sylvie and Bruno*. London: Macmillan, 1989.

Cartmill, Erica A., Sian Beilock, and Susan Goldin-Meadow. "A Word in the Hand: Action, Gesture, and Mental Representation in Humans and Nonhuman Primates." *Philosophical Transactions of the Royal Society B* 367 (2012): 129–43.

Cerling, Thure E., et al. "Woody Cover and Hominin Environments in the Past 6 Million Years." *Nature* 476 (2011): 51–56.

Cheney, Dorothy L., and Robert M. Seyfarth. *How Monkeys See the World*. Chicago: University of Chicago Press, 1990.

Chimpanzee Sequencing and Analysis Consortium. "Initial Sequence of the Chimpanzee Genome and Comparison with the Human Genome." *Nature* 437 (2005): 69–87.

Chomsky, Noam. *The Architecture of Language*. New Delhi: Oxford University Press, 2000.

———. "Biolinguistic Explorations: Design, Development, Evolution." *International Journal of Philosophical Studies* 15 (2007): 1–21.

———. *Derivation by Phase*. Cambridge, MA: MIT Press, 1999.

———. *New Horizons in the Study of Language and Mind*. Cambridge: Cambridge University Press. 2000.

———. "On Cognitive Structures and Their Development: A Reply to Piaget." In *Language and Learning: The Debate between Jean Piaget and Noam Chomsky*, edited by Massimo Piattelli-Palmarini, 35–52. Cambridge, MA: Harvard University Press, 1980.

———. *Reflections on Language*. New York: Pantheon, 1975.

———. "A Review of Skinner's 'Verbal Behavior.'" *Language* 35 (1959): 26–58.

———. "Some Simple Evo Devo Theses: How True Might They Be for Language?" In *The Evolution of Human Language*, edited by Richard K. Larson, Viviane Déprez, and Hiroko Yamakido, 45–62. Cambridge: Cambridge University Press, 2010.

———. *Syntactic Structures*. The Hague: Mouton, 1957.

Christiansen, Morten H., and Nick Chater. "Language as Shaped by the Brain." *Behavioral and Brain Sciences* 31 (2008): 489–558.

———. "Toward a Connectionist Model of Recursion in Human Linguistic Performance." *Cognitive Science* 23 (1998): 157–205.

Christiansen, Morten H., and Simon Kirby. "Language Evolution: The Hardest Problem in Science?" In *Language Evolution*, edited by Morten H. Christiansen and Simon Kirby, 1–15. Oxford: Oxford University Press, 2003.

< 233 >

Clayton, Nicola S., Timothy J. Bussey, and Anthony Dickinson. "Can Animals Recall the Past and Plan for the Future?" *Trends in Cognitive Sciences* 4 (2003): 685–91.

Collin, Silvy H. P., Branka Milivojevic, and Christian F. Doeller. "Memory Hierarchies Map onto the Hippocampal Long Axis in Humans." *Nature Neuroscience*, 2015, doi:10.1038/ nn.4138.

Colonnesi, Cristina, Geert Jan J. M. Stams, Irene Koster, and Marc J. Noomb. "The Relation between Pointing and Language Development: A Meta-analysis." *Developmental Review* 30 (2010): 352–66.

Condillac, Étienne Bonnot de. *An Essay on the Origin of Human Knowledge: Being a Supplement to Mr. Locke's "Essay on the Human Understanding"* [1747]. Translated by T. Nugent, 1756. Repr., Gainesville, FL: Scholars' Facsimiles and Reprints, 1971.

Coop, Graham, Kevin Bullaughev, Francesca Luca, and Molly Przeworski. "The Timing of Selection of the Human FOXP2 gene." *Molecular Biology and Evolution* 25 (2008): 1257–9.

Corballis, Michael C. *From Hand to Mouth: The Origins of Language.* Princeton, NJ: Princeton University Press, 2002.

———. "The Gestural Origins of Language," *American Scientist* 87 (1999): 138–45.

———. "On the Evolution of Language and Generativity." *Cognition* 44 (1992): 197–226.

———. "The Origins of Modernity: Was Autonomous Speech the Critical Factor?" *Psychological Review* 111 (2004): 543–52.

———. *The Recursive Mind.* Princeton, NJ: Princeton University Press, 2011.

———. The Wandering Mind. Chicago: University of Chicago Press, 2015.

———. "The Word According to Adam: The Role of Gesture in Language Evolution." In *From Gesture in Conversation to Visible Action as Utterance: Essays in Honor of Adam Kendon*, edited by Mandana Seyfeddinipur and Marianne Gullberg, 177–97. Amsterdam: John Benjamins, 2014.

Corkin, Suzanne. *Permanent Present Tense: The Man with No Memory, and What He Taught the World.* London: Allen Lane, 2013.

Correia, Sergio P. C., Anthony Dickinson, and Nicola S. Clayton. "Western Scrub-Jays Anticipate Future Needs Independently of Their Current Motivational State." *Current Biology* 17 (2007): 856–61.

Crawford, Michael A., and C. Leigh Broadhurst. "The Role of Docosahexaenoic and the Marine Food Web as Determinants of Evolution and Hominid Brain Development: The Challenge for Human Sustainability." *Nutritional Health* 21 (2012): 17–39.

Critchley, McDonald. *The Language of Gesture.* London: Arnold, 1939.

———. *Silent Language.* London: Arnold, 1975.

Cronk, Lee. "Continuity, Displaced Reference, and Deception." *Behavioral and Brain Sciences* 27 (2004): 510–11.

Crow, Timothy J. "Sexual Selection, Timing, and an X-Y Homologous Gene: Did *Homo sapiens* Speciate on the Y Chromosome?" In *The Speciation of Modern* Homo Sapiens, edited by Timothy J. Crow, 197–216. Oxford: Oxford University Press, 2002.

Cysouw, Michael, Dan Dediu, and Steven Moran. Comment on "Phonemic Diversity Supports a Serial Founder Effect Model of Language Expansion from Africa." *Science* 332 (2012): 346–48.

Darwin, Charles. *The Descent of Man and Selection in Relation to Sex.* 2nd ed. New York: Appleton, 1871.

< 234 >

——. *The Expression of the Emotions in Man and Animals*. London: John Murray, 1872.

——. *On the Origin of Species by Means of Natural Selection*. London: John Murray, 1859.

David, Deirdre. *The Cambridge Companion to the Victorian Novel*. Cambridge: Cambridge University Press, 2001.

Dawkins, Richard. *The Blind Watchmaker: Why the Evidence of Evolution Reveals a Universe without Design*. New York: W. W. Norton, 1986.

——. *The God Delusion*. New York: Random House, 2006.

——. *The Selfish Gene*. 2nd rev. ed. Oxford: Oxford University Press, 1989.

Deacon, Terrence. *The Symbolic Species*. New York: W. W. Norton, 1997.

Deane, Paul D. *Grammar in Mind and Brain: Explorations in Cognitive Syntax*. Berlin: Mouton de Gruyter, 1992.

——. "Neurological Evidence for a Cognitive Theory of Syntax: Agrammatic Aphasia and the Specialization of Form Hypothesis." In *Cognitive Linguistics in the Redwoods: The Expansion of a New Paradigm in Linguistics*, edited by Eugene H. Casad, 55–115. Berlin: Mouton de Gruyter, 1995.

De Boer, Bart, and W. Tecumseh Fitch. "Computer Models of Vocal Tract Evolution: An Overview and Critique." *Adaptive Behaviour* 18 (2010): 36–47.

Dediu, Dan, and Stephen C. Levinson. "On the Antiquity of Language: The Reinterpretation of Neandertal Linguistic Capacities and Its Consequences." *Frontiers in Psychology* 4 (2013), article 397.

Dennis, Megan Y., et al. "Evolution of Human-Specific Neural SRGAP2 Genes by Incomplete Segmental Duplication." *Cell* 149 (2012): 1–11.

Derdikman, Dori, and May-Britt Moser. "A Dual Role for Hippocampal Relay." *Neuron* 65 (2010): 582–84.

De Waal, Frans B. M. "The Antiquity of Empathy." *Science* 336 (2012): 874–76.

Diamond, Jared. "Deep Relationships among Languages." *Nature* 476 (2011): 291–92.

Dien, Joseph. "Looking Both Ways through Time: The Janus Model of Lateralised Cognition." *Brain and Cognition* 67 (2008): 292–323.

Ding, Nai, et al. "Cortical Tracking of Hierarchical Linguistic Structures in Connected Speech." *Nature Neuroscience* 19 (2016), doi:10.1038/nn.4186.

Dobson, Seth. "Allometry of Facial Mobility in Anthropoid Primates: Implications for the Evolution of Facial Expression." *American Journal of Physical Anthropology* 138 (2009): 70–81.

Dobson, Seth D., and Chet C. Sherwood. "Correlated Evolution of Brain Regions Involved in Producing and Processing Facial Expressions in Anthropoid Primates." *Biology Letters* 7 (2011): 86–88.

Dobzhansky, Theodosius. "Nothing in Biology Makes Sense Except in the Light of Evolution." *American Biology Teacher* 35 (1973): 125–29.

Dolscheid, Sarah, Sabine Hunnius, Daniel Casasanto, and Asifa Majid. "Prelinguistic Infants Are Sensitive to Space-Pitch Associations Found across Cultures." *Psychological Science* 25 (2014): 1256–61.

Donald, Merlin. *Origins of the Modern Mind*. Cambridge, MA: Harvard University Press, 1991.

Dor, Daniel. *The Instruction of Imagination: Language as a Social Communication Technology*. New York: Oxford University Press, 2015.

Dorren, Gaston. *Lingo: A Language Spotter's Guide to Europe*. London: Profile Books, 2014.

< 235 >

Duarte, Marcos M. "Red Ochre and Shells: Clues to Human Evolution." *Trends in Ecology and Evolution* 29 (2014): 560–65.

Duff, Melissa C., and Sarah Brown-Schmidt. "The Hippocampus and the Flexible Use and Processing of Language." *Frontiers in Human Neuroscience* 6 (2012), article 69.

Duffy, Carol Ann. "The Laughter of Stafford Girls' High." In *Feminine Gospels*, 35–54. London: Picador, 2003.

Dunbar, Robin. "Bridging the Bonding Gap: The Transition from Primates to Humans." *Proceedings of the Royal Society B* 367 (2012): 1837–46.

——. "Coevolution of Neocortical Size, Group Size, and Language in Humans." *Behavioral and Brain Sciences* 16 (1993): 681–735.

——. *The Human Story.* London: Faber and Faber, 2004.

——. "The Social Brain: Mind, Language and Society in Evolutionary Perspective." *Annual Review of Anthropology* 32 (2003): 163–81.

Dunbar, Robin, and John A. Gowlett. "Fireside Chat: The Impact of Fire on Hominin Socioecology." In *Lucy to Language: The Benchmark Papers*, edited by Robin Dunbar, Clive Gamble, and John Gowlett, 277–96. Oxford: Oxford University Press, 2014.

Einstein, Albert. *The World as I See It.* Secaucus, NJ: Citadel, 1979.

Emmorey, Karen. *Language, Cognition, and Brain: Insights from Sign Language Research.* Hillsdale, NJ: Erlbaum, 2002.

Enard, Wolfgang, et al. "A Humanized Version of Foxp2 Affects Cortico-Basal Ganglia Circuits in Mice." *Cell* 137 (2009): 961–71.

The Epic of Gilgamesh: The Babylonian Epic Poem and Other Texts in Akkadian and Sumerian. London: Penguin, 2000.

Evans, Nicholas. *Dying Words: Endangered Languages and What They Have to Tell Us.* Oxford: Wiley-Blackwell, 2009.

Evans, Nicholas, and Stephen C. Levinson. "The Myth of Language Universals: Language Diversity and Its Importance for Cognitive Science." *Behavioral and Brain Sciences* 32 (2009): 429–92.

Evans, Vyvyan. "Cognitive Linguistics." *WIRES Cognitive Science* 3 (2012): 129–41.

——. *The Language Myth: Why Language Is Not an Instinct.* Cambridge: Cambridge University Press, 2004.

Everett, Daniel L. "Cultural Constraints on Grammar and Cognition in Pirahã." *Current Anthropology* 46 (2005): 621–46.

——. *Language: The Cultural Tool.* New York: Pantheon, 2012.

Fadiga, Luciano, Leonardo Fogassi, Giovanni Pavesi, and Giacomo Rizzolatti. "Motor Facilitation during Action Observation: A Magnetic Stimulation Study." *Journal of Neurophysiology* 73 (1995): 2608–11.

Fano, Giorgio. *The Origins and Nature of Language.* Bloomington, IN: Indiana University Press, 1992. (Originally published in two parts, first in 1962 and second in 1973, and translated by S. Petrilli.)

Fauconnier, Gilles. "Cognitive Linguistics." In *Encyclopedia of Cognitive Science*, edited by Lynn Nadel, 1:539–43. London: Nature Publishing Group, 2003.

Fell, Juergen. "I Think, Therefore I Am (Unhappy)." *Frontiers in Human Neuroscience* 6 (2012), article 132.

Ferrer i Cancho, Ramon, and Ricard V. Solé. "Least Effort and the Origins of Scaling in

< 236 >

Human Language." *Proceedings of the National Academy of Sciences (USA)* 100 (2003): 788–91.

Fisher, Simon E., et al. "Localisation of a Gene Implicated in a Severe Speech and Language Disorder," *Nature Genetics* 18 (1998): 168–70.

Fitch, W. Tecumseh. *The Evolution of Language*. Cambridge: Cambridge University Press, 2010.

Fodor, Jerry A. *The Language of Thought*. New York: Crowell, 1975.

Foley, Robert A. "Early Man and the Red Queen: Tropical African Community Evolution and Hominid Adaptation." In *Hominid Evolution and Community Ecology: Prehistoric Human Adaptation in Biological Perspective*, edited by Robert A. Foley, 85–110. London: Academic, 1984.

Fonagy, Peter, and Mary Target. "The Rooting of the Mind in the Body: New Links between Attachment Theory and Psychoanalytic Thought." *Journal of the American Psychoanalytic Association* 55 (2007): 411–56.

Frame, Janet. *An Autobiography*. Collected ed. Auckland, NZ: Century Hutchinson, 1989.

Frank, Frederick S., and Anthony Magistrale. *The Poe Encyclopedia*. Westwood, CT: Greenwood.

Frishberg, Nancy. "Arbitrariness and Iconicity in American Sign Language." *Language* 51 (1975): 696–719.

Frisk, Virginia, and Brenda Milner. "The Role of the Left Hippocampal Region in the Acquisition and Retention of Story Content." *Neuropsychologia* 28 (1990): 349–59.

Fritz, Jan, Mortimer Mishkin, and Richard C. Saunders. "In Search of an Auditory Engram." *Proceedings of the National Academy of Sciences USA* 102 (2005): 9,359–64.

Fu, Qiaomei, et al. "An Early Modern Human from Romania with a Recent Neanderthal Ancestor." *Nature* 524 (2015): 216–19.

Gallese, Vittorio, et al. "Mirror Neuron Forum." *Perspectives on Psychological Science* 6 (2011): 369–407.

Ganis, Giorgio, and Haline E. Schendan. "Visual Imagery." *WIRES Cognitive Science* 2 (2011): 239–52.

Gärdenfors, Peter, and Mathias Osvath. "Prospection as a Cognitive Precursor to Symbolic Communication." In *The Evolution of Human Language*, edited by Richard K. Larson, Viviane Déprez, and Hiroko Yamakido, 103–14. Cambridge: Cambridge University Press, 2010.

Gardner, R. Allen, and Beatrice T. Gardner. "Teaching Sign Language to a Chimpanzee." *Science* 165 (1969): 664–72.

Gell-Mann, Murray, and Merritt Ruhlen. "The Origin and Evolution of Word Order." *Proceedings of the National Academy of Sciences USA* 108 (2011): 17,290–95.

Gentilucci, Maurizio, Francesca Benuzzi, Massimo Gangitano, and Silvia Grimaldi. "Grasp with Hand and Mouth: A Kinematic Study on Healthy Subjects." *Journal of Neurophysiology* 86 (2001): 1,685–99.

Gentilucci, Maurizio, and Michael C. Corballis. "From Manual Gesture to Speech: A Gradual Transition." *Neuroscience and Biobehavioral Reviews* 30 (2006): 949–60.

Gibbons, Ann. "A New Kind of Ancestor: *Ardipithecus* Unveiled." *Science* 326 (2009): 36–40.

Gibbs, Wolcott. "Time . . . Fortune . . . Life . . . Luce." *New Yorker*, November 28, 1936, 20–25.

Giebel, Ulrike, and D. Kimbrough Oller. "Vocabulary Learning in a Yorkshire Terrier: Slow Mapping of Spoken Words." *PLoS ONE* 7 (2012), article 30182.

< 237 >

Goldberg, Adele E. *Constructions: A Construction Grammar Approach to Argument Structure*. Chicago: University of Chicago Press, 1995.

Goodall, Jane. *The Chimpanzees of Gombe: Patterns of Behavior*. Cambridge, MA: Harvard University Press, 1986.

Goodwin, Ian D., Stuart A. Browning, and Athol J. Anderson. "Climate Windows for Polynesian Voyaging to New Zealand and Easter Island." *Proceedings of the National Academy of Sciences* 111:14,716–21.

Gopnik, Myrna. "Feature-Blind Grammar and Dysphasia." *Nature* 344 (1990): 715.

Gould, Stephen J. "Panselectionist Pitfalls in Parker and Gibson's Model of the Evolution of Intelligence." *Behavioral and Brain Sciences* 2 (1979): 385–86.

———. *Time's Arrow, Time's Cycle*. Cambridge, MA: Harvard University Press, 1987.

Gould, Stephen J., and Niles Eldredge. "Punctuated Equilibria: The Tempo and Mode of Evolution Reconsidered." *Paleobiology* 3 (1977): 115–51.

Gould, Stephen J., and Richard C. Lewontin. "The Spandrels of San Marco and the Panglossian Paradigm: A Critique of the Adaptationist Programme." *Proceedings of the Royal Society of London* 205 (1979): 281–88.

Gould, Stephen J., and Elisabeth S. Vrba. "Exaptation—a Missing Term in the Science of Form." *Paleobiology* 8 (1982): 4–15.

Gowlett, John A., and Richard W. Wrangham. "Earliest Fire in Africa: Towards the Convergence of Archeological Evidence and the Cooking Hypothesis." *Azania* 48 (2013): 5–30.

Green, Richard E., et al. "A Draft Sequence of the Neanderthal Genome." *Science* 328 (2010): 710–22.

Greene, Robert Lane. *You Are What You Speak*. New York: Delacorte, 2011.

Greenfield, Patricia, and E. Sue Savage-Rumbaugh. "Grammatical Combination in Pan paniscus: Processes of Learning and Invention in the Evolution and Development of Language." In "Language" and Intelligence in Monkeys and Apes, edited by Sue Taylor Parker and Kathleen Rita Gibson, 540–78. Cambridge: Cambridge University Press, 1990.

Grey, George. Polynesian Mythology and Ancient Traditional History of the New Zealand Race. Wellington, UK: A. H. and A. W. Reed, 1971. (Orig. London: John Murray, 1855; translated from George Grey, *Nga Mahi a Nga Tupuna* [1854]).

Grice, H. Paul. Studies in the Ways of Words. Cambridge: Cambridge University Press, 1989.

Griebel, Ulrike, and D. Kimbrough Oller. "Vocabulary Learning in a Yorkshire Terrier: Slow Mapping of Spoken Words." *PLoS ONE* 7 (2012): 30,182.

Griffin, Donald R. *Animal Minds: From Cognition to Consciousness*. Chicago: University of Chicago Press, 2001.

———. *The Question of Animal Awareness: Evolutionary Continuity of Mental Experience*. New York: Rockefeller University Press, 1976.

Gross, Charles G. "Huxley versus Owen: The Hippocampus Minor and Evolution." *Trends in Neuroscience* 16 (1993): 493–98.

Gupta, Anoopum S., et al. "Hippocampal Replay Is Not a Simple Function of Experience." *Neuron* 65 (2010): 695–705.

Haesler, Sebastian, et al. "Incomplete and Inaccurate Vocal Imitation after Knockdown of FOXP2 in Songbird Basal Ganglia Nucleus Area X." *PLoS Biology* 5 (2007): 2885–97.

Haldane, John B. S. "A Mathematical Theory of Natural and Artificial Selection, Part V: Selection and Mutation." *Proceedings of the Cambridge Philosophical Society* 23 (1927): 838–44.

< 238 >

Hammerschmidt, Kurt, et al. "A Humanized Version of Foxp2 Does Not Affect Ultrasonic Vocalization in Adult Mice." *Genes, Brain and Behavior* 14 (2015), doi 10.1111/gbb.12237.

Harari, Yuval N. *Sapiens: A Brief History of Humankind*. London: Harvill Secker, 2014.

Hare, Brian, Josep Call, Bryan Agnetta, and Michael Tomasello. "Chimpanzees Know What Conspecifics Do and Do Not See." *Animal Behaviour* 59 (2000): 771–85.

Hare, Brian, Josep Call, and Michael Tomasello. "Chimpanzees Deceive a Human Competitor by Hiding." *Cognition* 101 (2006): 495–514.

———. "Do Chimpanzees Know What Conspecifics Know?" *Animal Behaviour* 61 (2001): 139–51.

Harlaar, Nicole, et al. "Predicting Individual Differences in Reading Comprehension: A Twin Study." *Annals of Dyslexia* 60 (2010): 265–88.

Hartley, Leslie Poles. *The Go-Between*. London: Hamish Hamilton, 1953.

Hassabis, Demis, Dharshan Kumaran, and Eleanor A. Maguire. "Using Imagination to Understand the Neural Basis of Episodic Memory." *Journal of Neuroscience* 27 (2007): 14,365–74.

Hauser, Marc D., Noam Chomsky, and W. Tecumseh Fitch. "The Faculty of Language: What Is It, Who Has It, and How Did It Evolve?" *Science* 298 (2010): 1,569–79.

Hayes, Brian. "Belle Lettres Meets Big Data." *American Scientist* 102 (2014): 262–65.

Hayes, Cathy. *The Ape in Our House*. London: Gollancz, 1952.

Henn, Brenna M., Luigi Luca Cavalli-Sforza, and Marcus W. Feldman. "The Great Human Expansion." *Proceedings of the National Academy of Sciences USA* 109 (2012): 17,758–64.

Hepach, Robert, Amrisha Vaish, and Michael Tomasello. "Young Children Are Intrinsically Motivated to See Others Helped." *Psychological Science* 23 (2012): 967–72.

Hepper, Peter G., Glenda R. McCartney, and E. Alyson Shannon. "Lateralised Behaviour in First Trimester Human Fetuses." *Neuropsychologia* 36 (1998): 531–34.

Hepper, Peter G., Sara Shahidullah, and Raymond White. "Handedness in the Human Fetus." *Neuropsychologia* 29 (1991): 1101–11.

Hewes, Gordon W. "Primate Communication and the Gestural Origins of Language." *Current Anthropology* 14 (1973): 5–24.

Hickok, Gregory. "Eight Problems for the Mirror Neuron Theory of Action Understanding in Monkeys and Humans." *Journal of Cognitive Neuroscience* 21 (2009): 1229–43.

———. *The Myth of Mirror Neurons*. New York: W.W. Norton, 2014.

Hickok, Gregory, and David Poeppel. "The Cortical Organization of Speech Processing," *Nature Reviews Neuroscience* 8 (2007): 393–402.

Hinton, Geoffrey, et al. "Deep Neural Networks for Acoustic Modeling in Speech Recognition." *IEEE Signal Processing Magazine* 29 (2012): 82–97.

Hitchens, Christopher. *God Is Not Great: How Religion Poisons Everything*. New York: Little, Brown, 2007.

Hobaiter, Cat, and Richard W. Byrne. "The Meaning of Chimpanzee Gestures." *Current Biology* 24 (2014): 1–5.

———. "Serial Gesturing by Wild Chimpanzees: Its Nature and Function for Communication." *Animal Cognition* 14 (2011): 827–38.

Hobbes, Thomas. *Of Man*. Orig 1651; repr., Hoboken, NJ: Bibiobytes, 1999.

Hobson, J. Allan. "REM Sleep and Dreaming: Towards a Theory of Protoconsciousness." *Nature Reviews Neuroscience* 10 (2009): 803–14.

Hockett, Charles F. "In Search of Love's Brow." *American Speech* 53 (1978): 243–315.

< 239 >

Hoffecker, John F. "Innovation and Technological Knowledge in the Upper Paleolithic." *Evolutionary Anthropology* 14 (2005: 186–98.

———. "Representation and Recursion in the Archaeological Record." *Journal of Archaeological Method and Theory* 14 (2007): 359–87.

Hopkins, William D., Jared P. Taglialatela, and David A. Leavens. "Chimpanzees Differentially Produce Novel Vocalizations to Capture the Attention of a Human." *Animal Behaviour* 73 (2007): 281–86.

Hopper, Paul J., and Elizabeth Closs Traugott. *Grammaticalization*. 2nd ed. Cambridge: Cambridge University Press, 2003.

Horner, Victoria, J. Devyn Carter, Malini Suchak, and Frans B. M. de Waal. "Spontaneous Prosocial Choice by Chimpanzees." *Proceedings of the National Academy of Sciences USA* 108 (2011): 13,847–51.

Humphrey, Nicholas. "The Social Function of Intellect." In *Growing Points in Ethology*, edited by Patrick P. G. Bateson and Robert A. Hinde, 303–17. Cambridge: Cambridge University Press, 1976.

Hurford, James R. *The Origins of Grammar: Language in the Light of Evolution II*. Oxford: Oxford University Press, 2012.

———. *The Origins of Meaning: Language in the Light of Evolution*. Oxford: Oxford University Press, 2007.

Jackendoff, Ray. *Foundations of Language: Brain, Meaning, Grammar, Evolution*. Oxford: Oxford University Press, 2002.

———. "What Is the Human Language Faculty? Two Views." *Language* 87 (2011): 586–624.

James, William. *Principles of Psychology*. New York: Henry Holt, 1890.

Johansson, Sverker. "The Talking Neandertals: What Do Fossils, Genetics, and Archeology Say?" *Biolinguistics* 7 (2013): 35–74.

Jürgens, Uwe. "Neural Pathways Underlying Vocal Control." *Neuroscience and Biobehavioral Reviews* 26 (2002): 235–58.

Kamil, Alan C., and Russell P. Balda. "Cache Recovery and Spatial Memory in Clark's Nutcrackers (*Nucifraga columbiana*)." *Journal of Experimental Psychology: Animal Behavior Processes* 85 (1985): 95–111.

Kaminski, Juliane, Josep Call, and Julia Fischer. "Word Learning in the Domestic Dog: Evidence for 'Fast Mapping.'" *Science* 304 (2004): 1682–83.

Kano, Fumihiro, and Satoshi Hirato. "Great Apes Make Anticipatory Looks Based on Long-Term Memory of Single Events." *Current Biology* 25 (2015): 1–5.

Kant, Immanuel, "Anthropologie." In *Gesammelte Schriften* (1798). Translated in "The Grammar of Reason: Hammann's Challenge to Kant." *Synthèse* 75 (1988): 251–80.

Kaplan, Gisela. "Pointing Gesture in a Bird—Merely Instrumental or a Cognitively Complex Behaviour?" *Current Zoology* 57 (2011): 453–67.

Karlsson, Fred. "Constraints on Multiple Center-Embedding of Clauses." *Journal of Linguistics* 43 (2007): 365–92.

Kegl, Judy, Ann Senghas, and Marie Coppola. "Creation through Contact: Sign Language Emergence and Sign Language Change in Nicaragua." In *Language Creation and Language Change: Creolization, Diachrony, and Development*, edited by M. DeGraff, 179–237. Cambridge, MA: Bradford Book, MIT Press, 1999.

Kellogg, Winthrop N., and Luella A. Kellogg. *The Ape and the Child: A Study of Early Environmental Influence upon Early Behavior*. New York: McGraw-Hill, 1933.

< 240 >

Kendon, Adam. "Vocalisation, Speech, Gesture, and the Language Origins Debate." *Gesture* 13 (2011): 349–70.

Kingsley, Charles. *The Water-Babies: A Fairy Tale for a Land Baby.* London: Macmillan, 1863.

Kingsley, Mary. *Travels in West Africa, Cong Française, Corisco, and Cameroons.* London: F. Cass, 1897.

Kirby, Simon, and James R. Hurford. "The Emergence of Linguistic Structure: An Overview of the Iterated Learning Model." In *Simulating the Evolution of Language*, edited by Angelo Cangelosi and Domenico Parisi, 121–48. London: Springer Verlag, 2002.

Klein, Richard G. "Out of Africa and the Evolution of Human Behavior." *Evolutionary Anthropology* 17 (2008): 267–81.

Klein, Stanley B. "The Complex Act of Projecting Oneself into the Future." *WIREs Cognitive Science* 4 (2013): 63–79.

Klein, Stanley B., Theresa E. Robertson, and Andrew W. Delton. "Facing the Future: Memory as an Evolved System for Planning Future Acts." *Memory and Cognition* 38 (2010): 13–22.

Klima, Stephanie, and Ursula Bellugi. *The Signs of Language.* Cambridge, MA: Harvard University Press, 1979.

Knott, Alistair. *Sensorimotor Cognition and Natural language Syntax.* Cambridge, MA: MIT Press, 2012.

Kohler, Evelyne, et al. "Hearing Sounds, Understanding Actions: Action Representation in Mirror Neurons." *Science* 297 (2002): 846–48.

Köhler, Wolfgang. *Gestalt Psychology.* New York: Liveright, 1929.

Krause, Johannes, et al. "The Derived FOXP2 Variant of Modern Humans Was Shared with Neandertals." *Current Biology* 17 (2007): 1,908–12.

Kundera, Milan. *Ignorance.* Translated by L. Asher. New York: HarperCollins, 2002.

Langford, Dale J., et al. "Social Modulation of Pain as Evidence for Empathy in Mice. *Science* 312 (2006): 1,967–70.

Leavens, David A., and Timothy P. Racine. "Joint Attention in Apes and Humans." *Journal of Consciousness Studies* 16 (2000): 240–67.

Lewis-Williams, David, and David Pearce. *Inside the Neolithic Mind: Consciousness, Cosmos, and the Realm of the Gods.* London: Thames and Hudson, 2005.

Liberman, Alvin M., Franklin S. Cooper, Donald P. Shankweiler, and Michael Studdert-Kennedy. "Perception of the Speech Code." *Psychological Review* 74 (1967): 431–61.

Liebal, Katja, Josep Call, and Michael Tomasello. "Use of Gesture Sequences in Chimpanzees." *American Journal of Primatology* 64 (2004): 377–96.

Lieberman, Daniel E. "Sphenoid Shortening and the Evolution of Modern Cranial Shape." *Nature* 393 (1998): 158–62.

———. *The Story of the Human Body: Evolution, Health and Disease.* New York: Pantheon, 2013.

Lieberman, Daniel E., Brandeis M. McBratney, and Gail Krovitz. "The Evolution and Development of Cranial Form in *Homo sapiens*." *Proceedings of the National Academy of Sciences* 99 (2002): 1,134–39.

Lieberman, Philip. "The Evolution of Human Speech." *Current Anthropology* 48 (2007): 39–46.

Liégeois, Frédérique, et al. "Language fMRI Abnormalities Associated with FOXP2 Gene Mutation." *Nature Neuroscience* 6 (2003): 1,230–37.

Lin, Jo-Wang. "Time in a Language without Tense: The Case of Chinese." *Journal of Semantics* 23 (2006): 1–53.

Liszkowski, Ulf, Marie Schäfer, Malinda Carpenter, and Michael Tomasello. "Prelinguistic

< 241 >

Infants, but Not Chimpanzees, Communicate about Absent Entities." *Psychological Science* 20 (2009): 654–60.

Locke, John. *An Essay concerning Human Understanding*. Orig. 1690; repr. London: Scolar Press, 1970.

Locke, John L., and Barry Bogin. "Language and Life History: A New Perspective on the Development and Evolution of Human Language." *Behavioral and Brain Sciences* 29 (2006): 259–325.

Loftus, Elizabeth, and Katherine Ketcham. *The Myth of Repressed Memory*. New York: St. Martin's, 1994.

Lovejoy, C. Owen, et al. "The Pelvis and Femur of *Ardipithecus Ramidus*: The Emergence of Upright Walking." *Science* 326 (2009): 71, 71e1–71e6.

Lu, Hanbing, et al. "Rat Brains Also Have a Default Mode Network." *Proceedings of the National Academy of Sciences (USA)* 109 (2012): 3,979–84.

Lyn, Heidi, et al. "Apes Communicate about Absent and Displaced Objects: Methodology Matters." *Animal Cognition* 17 (2014): 85–94.

MacNeilage, Peter F. *The Origin of Speech*. Oxford: Oxford University Press, 2008.

Maguire, Eleanor A., et al. "Navigation-Related Cells: Structural Change in the Hippocampi of Taxi Drivers." *Proceedings of the National Academy of Sciences (USA)* 97 (2000): 4,398–403.

Maguire, Eleanor A., Katherine Woollett, and Hugo J. Spiers. "London Taxi Drivers and Bus Drivers: A Structural MRI and Neuropsychological Analysis." *Hippocampus* 16 (2006): 1091–101.

Mar, Raymond A., et al. "Bookworms versus Nerds: Exposure to Fiction versus Nonfiction, Divergent Associations with Social Ability, and the Simulation of Fictional Social Worlds." *Journal of Research in Personality* 40 (2006): 694–712.

Marshall, Jane, et al. "Aphasia in a User of British Sign Language: Dissociation between Sign and Gesture." *Cognitive Neuropsychology* 21 (2004): 537–54.

Martin, Victoria C., Daniel L. Schacter, Michael C. Corballis, and Donna Rose Addis. "A Role for the Hippocampus in Encoding Simulations of Future Events." *Proceedings of the National Academy of Sciences* 108 (2011): 13,858–63.

McBrearty, Sally, and Alison S. Brooks. "The Revolution That Wasn't: A New Interpretation of the Origin of Modern Human Behavior." *Journal of Human Evolution* 39 (2000): 453–563.

McBride, Glen. "Story Telling, Behavior Planning, and Language Evolution in Context." *Frontiers in Psychology* 5 (2014), article 1131.

McCawley, James D. *Thirty Million Theories of Grammar*. Chicago: University of Chicago Press, 1982.

McGurk, Harry, and John MacDonald. "Hearing Lips and Seeing Voices." *Nature* 264 (1976): 746–48.

McNeill, David. *How Language Began: Gesture and Speech in Human Evolution*. Cambridge: Cambridge University Press, 2012.

———. "So You Think Gestures Are Nonverbal?" *Psychological Review* 92 (1985): 350–71.

Mehler, Jacques, et al. "A Precursor of Language Acquisition in Young Infants." *Cognition* 29 (1988): 143–78.

Mellars, Paul A. "Going East: New Genetic and Archaeological Perspectives on the Modern Colonization of Eurasia." *Science* 313 (2006): 796–800.

< 242 >

——. "The Impossible Coincidence: A Single-Species Model for the Origins of Modern Human Behavior in Europe." *Evolutionary Anthropology* 14 (2005): 22–27.

——. "Origins of the Female Image." *Nature* 459 (2009): 176–77.

Mendenhall, Thomas Corwin. "The Characteristic Curves of Composition." *Science* 9 (1887): 237–49.

Menzel, Charles R. "Unprompted Recall and Reporting of Hidden Objects by a Chimpanzee (*Pan Troglodytes*) after Extended Delays." *Journal of Comparative Psychology* 113 (1969): 426–34.

Meunier, Helene, Jacques Vauclair, and Jacqueline Fagard. "Human Infants and Baboons Show the Same Pattern of Handedness for a Communicative Gesture." *PLoS ONE* 7 (2012), e33559.

Miles, H. Lyn White, and Stephen E. Harper. "Ape Language Studies and the Study of Human Language Origins." In *Hominid Culture in Primate Perspective*, edited by Duane D. Quiatt and Junichiro Itani, 253–78. Niwot: University Press of Colorado, 1994.

Miller, David J., et al. "Prolonged Myelination in Human Neocortical Evolution." *Proceedings of the National Academy of Sciences (USA)* 109 (2012): 16,480–85.

Miller, Jonathan F., et al. "Neural Activity in Human Hippocampal Formation Reveals the Spatial Context of Retrieved Memories." *Science* 342 (2013): 1111–14.

Milne, A. A. [Alan Alexander]. *Now We Are Six*. London: Methuen, 1927.

Milner, A. David, and Melvyn A. Goodale. *The Visual Brain in Action*. 2nd ed. Oxford: Oxford University Press, 2006.

Mishkin, Mortimer, Leslie G. Ungerleider, and Kathleen A. Macko. "Object Vision and Spatial Vision: Two Cortical Pathways." *Trends in Neuroscience* 6 (1983): 414–17.

Molenberghs, Pascal, Ross Cunnington, and Jason B. Mattingley. "Brain Regions with Mirror Properties: A Meta-analysis of 125 Human FMRI Studies." *Neuroscience and Biobehavioral Reviews* 36 (2012): 341–49.

Morgan, Elaine. *The Aquatic Ape*. New York: Stein and Day, 1982.

Moser, May-Britt, David C. Rowland, and Edvard I. Moser. "Place Cells, Grid Cells, and Memory." *Cold Spring Harbor Perspectives in Biology* 7 (2015): a021808.

Moseley, Christopher, ed. *The UNESCO Atlas of the World's Languages in Danger: Content and Process*. Memory of Peoples Series, Occasional Paper 5. Cambridge: World Oral Literature Project, 2012.

Muir, Laura J., and Iain E. G. Richardson. "Perception of Sign Language and Its Application to Visual Communication for Deaf People." *Journal of Deaf Studies and Deaf Education* 10 (2005): 390–401.

Mukamel, Roy, et al. "Single-Neuron Responses in Humans during Execution and Observation of Actions." *Current Biology* 20 (2010): 750–56.

Mulcahy, Nicholas J., and Josep Call. "Apes Save Tools for Future Use." *Science* 312 (2006): 1038–40.

Mulcahy, Nicholas J., and Vernon Hedge. "Are Great Apes Tested with an Abject Object-Choice Task?" *Animal Behaviour* 83 (2012): 313–21.

Müller, Friedrich. "Lectures on Mr Darwin's Philosophy of Language." *Fraser's Magazine* –8 (1873). Repr. in *The Origin of Language*, edited by R. Harris, 147–233. Bristol, UK: Thoemmes, 1996.

Nelson, Katherine. *Language in Cognitive Development: The Emergence of the Mediated Mind*. Cambridge: Cambridge University Press, 1996.

< 243 >

Newell, Alan, John Calman Shaw, and Herbert A. Simon. "Elements of a Theory of Human Problem Solving." *Psychological Review* 23 (1958): 151–66.

Newton, Michael. *Savage Girls and Wild Boys: A History of Feral Children*. London: Faber and Faber, 2004.

Nietzsche, Friedrich W. *Human, All Too Human: A Book for Free Spirits* [1878]. Translated by R. J. Hollingdale. New York: Cambridge University Press, 1986.

Nikolaïdes, Kimon. *The Natural Way to Draw*. Boston: Houghton Mifflin, 1975.

Niles, John D. *Homo narrans: The Poetics and Anthropology of Oral Literature*. Philadelphia: University of Pennsylvania Press, 2010.

Noonan, James P., et al. "Sequencing and Analysis of Neanderthal Genomic DNA." *Science* 314 (2006): 1113–21.

O'Keefe, John. "The Spatial Prepositions in English, Vector Grammar, and the Cognitive Map Theory." In *Language and Space*, edited by Paul Bloom, Mary A. Peterson, Lynne Nadel, and Merrill F. Garrett, 277–316. Cambridge, MA: MIT Press, 1996.

O'Keefe, John, and Lynn Nadel. *The Hippocampus as a Cognitive Map*. Oxford: Clarendon, 1978.

Oatley, Keith. "The Cognitive Science of Fiction." *WIREs Cognitive Science* 3 (2012): 425–30.

Ohnuki-Tierney, Emiko. *Illness and Healing among the Sakhalin Ainu: A Symbolic Interpretation*. Cambridge: CUP Archive, 1981.

Orwell, George. *Animal Farm*. London: Secker and Warburg, 1945.

Osvath, Mathias, and Elin Karvonen. "Spontaneous Innovation for Future Deception in a Male Chimpanzee." *PLoS ONE* 7 (2012), e36782.

Paget, Richard A. S. "The Origin of Speech—a Hypothesis." *Proceedings of the Royal Society of London A*, 119 (1928): 157–72.

Pande, Hemlata, and Hoshiya S. Dhami. "Model Generation for Word Length Frequencies in Texts with the Application of Zipf's Order Approach." *Journal of Quantitative Linguistics* 19 (2012): 249–61.

Pastalkova, Eva, Vladimir A. Itskov, and György Buzsáki. "Internally Generated Cell Assembly Sequences in the Rat Hippocampus." *Science* 321 (2008): 1322–27.

Pastra, Katerina, and Yiannis Aloimonos. "The Minimalist Grammar of Action." *Philosophical Transactions of the Royal Society B* 367 (2012): 103–17.

Patterson, Francine G. P., and Wendy Gordon. "Twenty-Seven Years of Project Koko and Michael." In *All Apes Great and Small*, vol. 1, *African Apes*, edited by Biruté Galdikas, N. Erickson Briggs, Lori K. Sheeran, and Jane Goodall, 165–76. New York: Kluver, 2001.

Penn, Derek C., Keith J. Holyoak, and Daniel J. Povinelli. "Darwin's Mistake: Explaining the Discontinuity between Human and Nonhuman Minds." *Behavioral and Brain Sciences* 31 (2008): 108–78.

Pennisi, Elizabeth. "The Burdens of Being a Biped." *Science* 336 (2012): 974.

Perniss, Pamela, and Gabriella Vigliocco. "The Bridge of Iconicity: From a World of Experience to the Experience of Language." *Philosophical Transactions of the Royal Society B* 369 (2014), 20130300.

Petkov, Christopher I., and Erich D. Jarvis. "Birds, Primates, and Spoken Language Origins: Behavioral Phenotypes and Neurobiological Substrates." *Frontiers in Evolutionary Neuroscience* 4 (2012), article 12.

Petrides, Michael, and Deepak N. Pandya. "Distinct Parietal and Temporal Pathways to the Homologue of Broca's Area in the Monkey." *PLoS ONE* 7 (8) (2009), e1000170.

< 244 >

Pettito, Laura A., and Pauline Marentette. "Babbling in the Manual Mode: Evidence for the Ontogeny of Language." *Science* 251 (1991): 1493–96.

Pfeiffer, John E. *The Emergence of Man*. London: Book Club Associates, 1973.

Pfenning, Andreas R., et al. "Convergent Specializations in the Brains of Humans and Song-Learning Birds." *Science* 346 (2014): 1333–46.

Pietrandrea, Paola. "Iconicity and Arbitrariness in Italian Sign Language." *Sign Language Studies* 2 (2002): 296–321.

Piff, Paul K., et al. "Higher Social Class Predicts Increased Unethical Behavior." *Proceedings of the National Academy of Sciences USA* 109 (2012): 4087–91.

Pika, Simone, Katja Liebal, and Michael Tomasello. "Gestural Communication in Subadult Bonobos (*Pan paniscus*): Repertoire and Use." *American Journal of Primatology* 65 (2005): 39–61.

———. "Gestural Communication in Young Gorillas (*Gorilla gorilla*): Gestural Repertoire and Use." *American Journal of Primatology* 60 (2003): 95–111.

Pika, Simone, and John C. Mitani. "The Directed Scratch: Evidence for a Referential Gesture in Chimpanzees?" In *The Prehistory of Language*, edited by Rudolf Botha and Chris Knight. Oxford: Oxford Scholarship Online, 2009; doi:10.1093/acprof:oso/9780199545872.003.0009.

Pilley, John W., and Alliston K. Reid. "Border Collie Comprehends Object Names as Verbal Referents." *Behavioural Processes* 86 (2011): 184–95.

Pinhasi, Ron, Thomas F. G. Higham, Liubov Golovanova, and Vladimir Doronichev. "Revised Age of Late Neanderthal Occupation and the End of the Middle Paleolithic in the Northern Caucasus." *Proceedings of the National Academy of Sciences* 108 (2011): 8611–16.

Pinker, Steven. "Language as an Adaptation to the Cognitive Niche." In *Language Evolution*, edited by Morten H. Christiansen and Simon Kirby, 16–37. Oxford: Oxford University Press, 2003.

———. *The Language Instinct*. New York: William Morrow, 1994.

———. *The Stuff of Thought*. London: Penguin Books, 2007.

Pinker, Steven, and Paul Bloom. "Natural Language and Natural Selection." *Behavioral and Brain Sciences*, 13 (1990): 707–84.

Pizzuto, Elena, and Virginia Volterra. "Iconicity and Transparency in Sign Languages: A Cross-Linguistic Cross-Cultural View." In *The Signs of Language Revisited: An Anthology to Honor Ursula Bellugi and Edward Klima*, edited by Karen Emmorey and Harlan Lane, 261–86. Mahwah, NJ: Lawrence Erlbaum Associates, 2000.

Plato. *The Dialogues of Plato*. Translated by B. Jowett. Vol. 4. Oxford,: Oxford University Press, 1892.

Ploog, Detlev. "Is the Neural Basis of Vocalisation Different in Non-human Primates and *Homo sapiens*?" In *The Speciation of Modern Homo Sapiens*, edited by Timothy J. Crow, 121–35. Oxford: Oxford University Press, 2002.

Plooij, Frans X. "Some Basic Traits of Language in Wild Chimpanzees?" In *Action, Gesture and Symbol: The Emergence of Language*, edited by A. Lock, 111–31. New York: Academic, 1978.

Plutarch. "Life of Pompeius." In *Plutarch's Lives*, edited by Aubrey Stewart and George Long, translated by George Long, 195–294. London: George Bell and Sons, 1892.

Poizner, Howard, Edward S. Klima, and Ursula Bellugi. *What the Hands Reveal about the Brain*. Cambridge, MA: MIT Press, 1984.

< 245 >

Pollick, Amy S., and Frans B. M. de Waal. "Apes Gestures and Language Evolution." *Proceedings of the National Academy of Sciences* 104 (2007): 81–89.

Poole, Jean, and Denis G. Lander. "The Pigeon's Concept of Pigeon." *Psychonomic Science* 25 (1971): 157–58.

Povinelli, Daniel J., Jesse M. Bering, and Steve Giambrone. "Toward a Science of Other Minds: Escaping the Argument by Analogy." *Cognitive Science* 24 (2000): 509–41.

Premack, David. "Gavagai! or The Future History of the Animal Language Controversy." *Cognition* 19 (1985): 207–96.

———. "Human and Animal Cognition: Continuity and Discontinuity." *Proceedings of the National Academy of Sciences (USA)* 104 (2007): 13,861–67.

Premack, David, and Guy Woodruff. "Does the Chimpanzee Have a Theory of Mind?" *Behavioral and Brain Sciences* 4 (1978): 515–26.

Provine, Robert. *Laughter: A Scientific Investigation.* London: Viking Penguin, 2001.

Pruetz, Jill D., and Paco Bertolani. "Savanna Chimpanzees, *Pan troglodytes verus*, Hunt with Tools." *Current Biology* 17 (2007): 412–17, http://www.smithsonianmag.com/science-nature/speakingbonobo.html.

Ptak, Susan E., et al. "Linkage Disequilibrium Extends across Putative Selected Sites in FOXP2." *Molecular and Biological Evolution* 26 (2012): 2181–84.

Pylyshyn, Zenon W. "What the Mind's Eye Tells the Mind's Brain: A Critique of Mental Imagery." *Psychological Bulletin* 80 (1973): 1–24.

Quilliam, Susan. "'He Seized Her in His Manly Arms and Bent His Lips to Hers . . .': The Surprising Impact That Romantic Novels Have on Our Work." *Journal of Family Planning, Reproduction and Health Care* 37 (2011): 179–81.

Raichle, Marcus E., et al. "A Default Mode of Brain Function." *Proceedings of the National Academy of Sciences* 109 (2001): 3979–84.

Ramachandran, Vilayanur S., and Edward M. Hubbard. "Synaesthesia—A Window into Perception, Thought and Language." *Journal of Consciousness Studies* 8 (2001): 3–34.

Rand, David G., Joshua D. Greene, and Martin A. Nowak. "Spontaneous Giving and Calculated Greed." *Nature* 489 (2012): 427–30.

Reich, David, et al. "Genetic History of an Archaic Hominin Group from Denisova Cave in Siberia." *Nature* 468 (2010): 1053–60.

Rivas, Esteban. "Recent Use of Signs by Chimpanzees (*Pan troglodytes*) in Interactions with Humans." *Journal of Comparative Psychology* 119 (2005): 404–17.

Rizzolatti, Giacomo, and Michael A. Arbib. "Language within Our Grasp." *Trends in Neurosciences* 21 (1998): 188–94.

Rizzolatti, Giacomo, et al. "Functional Organization of Inferior Area 6 in the Macaque Monkey, Part 2, Area F5 and the Control of Distal Movements." *Experimental Brain Research* 71 (1988): 491–507.

Rizzolatti, Giacomo, and Corrado Sinigaglia. "The Functional Role of the Parieto-Frontal Mirror Circuit: Interpretations and Misinterpretations." *Nature Neuroscience* 11 (2010): 264–74.

———. *Mirrors in the Brain: How Our Minds Share Actions and Emotions.* Oxford: Oxford University Press, 2006.

Rodríguez-Vidal, Joaquin, et al. "A Rock Engraving Made by Neanderthals in Gibraltar." *Proceedings of the National Academy of Sciences* 111 (2014): 13,301–6.

< 246 >

Rousseau, Jean-Jacques. *Essay on the Origin of Languages* [1782]. Translated by John H. Moran and Alexander Gode. Chicago: University of Chicago Press, 1966.

Rovelli, Carlo. *Seven Brief Lessons on Physics*. London: Penguin Books, 2015.

Ruhlen, Merritt. *The Origin of Language: Tracing the Origin of the Mother Tongue*. New York: Wiley, 1994.

Russell, Bailey A., Frank J. Cerny, and Elaine T. Stathopoulos. "Effects of Varied Vocal Intensity on Ventilation and Energy Expenditure in Women and Men." *Journal of Speech, Language, and Hearing Research* 41 (1998): 239–48.

Russon, Anne, and Kristin Andrews. "Orangutan Pantomime: Elaborating the Message." *Biology Letters* 7 (2001): 627–30.

Salmond, Ann. "Mana Makes the Man: A Look at Maori Oratory and Politics." In *Political Language and Oratory in Traditional Society*, edited by M. Bloch, 45–63. New York: Academic, 1975.

Saussure, Ferdinand de. *Course in General Linguistics* [*Cours de linguistique générale*, 1916]. Translated by W. Baskin. Glasgow: Fontana/Collins, 1977.

Savage-Rumbaugh, Sue, Stuart G. Shanker, and Talbot J. Taylor. *Apes, Language, and the Human Mind*. New York: Oxford University Press, 1998.

Schultze-Berndt, Eva. *Simple and Complex Verbs in Jaminjung: A Study of Event Categorization in an Australian Language*. PhD diss., Radboud University, MPI Series in Psycholinguistics, 2000.

Schulze, Katrin, Faraneh Vargha-Khadem, and Mortimer Mishkin. "Test of a Motor Theory of Long-Term Auditory Memory." *Proceedings of the National Academy of Sciences* 109 (2012): 7121–25.

Scott-Phillips, Thomas C. *Speaking Our Minds: Why Human Communication Is Different, and How Language Evolved to Make It Special*. Basingstoke, UK: Palgrave Macmillan, 2015.

Senghas, Ann, Sotaro Kita, and Asli Özyürek. "Children Creating Core Properties of Language: Evidence from an Emerging Sign Language in Nicaragua." *Science* 305 (2004): 1780–82.

Seyfarth, Robert M., Dorothy L. Cheney, and Peter Marler. "Monkey Responses to Three Different Alarm Calls—Evidence of Predator Classification and Semantic Communication." *Science* 210 (1980): 801–3.

Shea, John J. "*Homo sapiens* Is as *Homo sapiens* Was." *Current Anthropology* 52 (2011): 1–35.

Shintel, Hadas, and Howard C. Nusbaum. "The Sound of Motion in Spoken Language: Visual Information Conveyed by Acoustic Properties of Speech." *Cognition* 105 (2007): 681–90.

Shintel, Hadas, Howard C. Nusbaum, and Arika Okrent. "Analog Acoustic in Speech." *Journal of Memory and Language* 55 (2006): 167–77.

Shu, Weiguo, et al. "Characterization of a New Subfamily of Winged-Helix/Forkhead (Fox) Genes That Are Expressed in the Lung and Act as Transcriptional Repressors." *Journal of Biological Chemistry* 276 (2001): 27,488–97.

Skinner, Burrhus F. *Verbal Behavior*. New York: Appleton-Century-Crofts, 1957.

Skyrms, Brian. *The Stag Hunt and the Evolution of Social Structure*. Cambridge: Cambridge University Press, 2003.

Slocombe, Katie E., and Klaus Zuberbühler. "Chimpanzees Modify Recruitment Screams as a Function of Audience Composition." *Proceedings of the National Academy of Sciences* (USA) 104 (2007): 17,228–33.

< 247 >

Slobin, Dan I. "From 'Thought and Language' to 'Thinking for Speaking.'" In *Rethinking Linguistic Relativity*, edited by John Gumperz and Stephen C. Levinson, 70–96. Cambridge: Cambridge University Press, 1960.

Sosis, Richard. "The Adaptive Value of Religious Ritual." *American Scientist* 92 (2004): 166–72.

Spelke, Elizabeth S., and Katherine D. Kinzler. "Core Knowledge." *Developmental Science* 10 (2007): 89–96.

Sperber, Dan, and Deirdre Wilson. "Pragmatics, Modularity and Mind-Reading." *Mind and Language* 17 (2002): 3–23.

Spreng, R. Nathan, Raymond A. Mar, and Alice S. N. Kim. "The Common Neural Basis of Autobiographical Memory, Prospection, Navigation, Theory of Mind, and the Default Mode: A Quantitative Meta-analysis." *Journal of Cognitive Neuroscience* 21 (2009): 489–510.

Stelma, Juurd H., and Lynne J. Cameron. "Intonation Units in Spoken Interaction: Developing Transcription Skills." *Text and Talk* 27 (2007): 361–93.

Sterelny, Kim. "Social Intelligence, Human Intelligence, and Niche Construction." *Philosophical Transactions of the Royal Society B* 362 (2007): 719–30.

Stokoe, William C. *Language in Hand: Why Sign Came before Speech*. Washington, DC: Gallaudet University Press, 2001.

Stout, Dietrich, and Thierry Chaminade. "Stone Tools, Language and the Brain in Human Evolution." *Philosophical Transactions of the Royal Society B: Biological Sciences* 367 (2012): 75–87.

Strange, Bryan A., Menno P. Witter, Ed S. Lein, and Edvard I. Moser. "Functional Organization of the Hippocampal Longitudinal Axis." *Nature Reviews Neuroscience* 15 (2014): 655–69.

Suddendorf, Thomas. *The Gap: The Science of What Separates Us from Other Animals*. New York: Basic Books, 2013.

Suddendorf, Thomas, and Michael C. Corballis. "Mental Time Travel and the Evolution of the Human Mind." *Genetic, Social, and General Psychology Monographs* 123 (1997): 133–67.

Sutherland, John. *A Little History of Literature*. New Haven, CT: Yale University Press, 2013.

Sutton-Spence, Rachel, and Penny Boyes-Braem, eds. *The Hands Are the Head of the Mouth: The Mouth as Articulator in Sign Language*. Hamburg: Signum-Verlag, 2001.

Szathmàry, Eörs. "Toward Major Evolutionary Transitions 2.0." *Proceedings of the National Academy of Sciences (USA)* 112 (2015): 10,104–11.

Tanner, Joanne E., Francine G. Patterson, and Richard W. Byrne. "The Development of Spontaneous Gestures in Zoo-Living Gorillas and Sign-Taught Gorillas: From Action and Location to Object Representation." *Journal of Developmental Processes* 1 (2006): 69–103.

Tattersall, Ian. *Masters of the Planet: The Search for Human Origins*. New York: Palgrave Macmillan, 2012.

Thieme, Hartmut. "Lower Palaeolithic Hunting Spears from Germany." *Nature* 385 (1997): 807–10.

Thorpe, Suzanna K. S., Roger L. Holder, and Robin H. Crompton. "Origin of Human Bipedalism as an Adaptation for Locomotion on Flexible Branches." *Science* 316 (2007): 1328–31.

Tobias, Philip V. "Revisiting Water and Hominin Evolution." In *Was Man More Aquatic in the Past? Fifty Years after Alister Hardy*, edited by Mario Vaneechoutte, Algis Kuliakis, and Marc Verhaegen, 3–15. Oak Park, IL: Bentham Science, 2011.

< 248 >

Tomasello, Michael. *The Origins of Human Communication*. Cambridge, MA: MIT Press, 2008.

———. "Universal Grammar Is Dead." *Behavioral and Brain Sciences* 32 (2009): 470.

Tooby, John, and Irven DeVore. "The Reconstruction of Hominid Evolution through Strategic Modeling." In *The Evolution of Human Behavior: Primate Models*, edited by W. G. Kinzey, 183–237. Albany: SUNY Press, 1987.

Tulving, Endel. "Episodic and Semantic Memory." In *Organization of Memory*, edited by Endel Tulving and Wayne Donaldson, 381–403. New York: Academic, 1972.

———. "Memory and Consciousness." *Canadian Psychologist* 26 (1985): 1–12.

Turin, Mark. "Voices of Vanishing Worlds: Endangered Languages, Orality and Cognition." *Analise Social* 47 (2012): 846–69.

Urban, Matthias. "Conventional Sound Symbolism in Terms for Organs of Speech: A Cross-Linguistic Study." *Folia Linguistica* 45 (2011): 199–214.

Vargha-Khadem, Faraneh, et al. "Praxic and Nonverbal Cognitive Deficits in a Large Family with a Genetically Transmitted Speech and Language Disorder." *Proceedings of the National Academy of Sciences (USA)* 92 (1995): 930–33.

Verhaegen, Mark. "The Aquatic Ape Evolves: Common Misconceptions and Unproven Assumptions about the So-Called Aquatic Ape Hypothesis." *Human Evolution* 28 (2013): 237–66.

———. "Origin of Hominid Bipedalism." *Nature* 325 (1987): 305–6.

Villa, Paolo, and Wil Roebroeks. "Neandertal Demise: An Archaeological Analysis of the Modern Human Superiority Complex." *PLoS ONE* 9 (2014), e96424.

Villmoore, Brian, et al. "Early Homo at 2.8 Ma from Ledi-Geraru, Afar, Ethiopia." *Science Express*, March 5, 2015, doi 10.1126/science.aaa1343.

Vincent, Justin L., et al. "Intrinsic Functional Architecture in the Anaesthetized Monkey Brain." *Nature* 447 (2007): 83–86.

Volterra, Virginia, Maria Cristina Caselli, Olga Capirci, and Elena Pizzuto. "Gesture and the Emergence and Development of Language." In *Beyond Nature-Nurture: Essays in Honor of Elizabeth Bates*, edited by Michael Tomasello and Dan Slobin, 3–40. Mahwah, NJ: Lawrence Erlbaum Associates, 2005.

von Frisch, Karl. *The Dance Language and Orientation of Bees*. Cambridge, MA: Harvard University Press, 1967.

Vygotsky, Lev. *Mind in Society*. Cambridge, MA: Harvard University Press, 1978.

Wallace, Alfred Russel. "Sir Charles Lyell on Geological Climates and the Origin of Species." *Quarterly Review* 126 (1869): 359–94.

Wearing, Deborah. *Forever Today*. New York: Doubleday, 2005.

Wechkin, Stanley, Jules H. Masserman, and William Terris. "Shock to a Conspecific as an Aversive Stimulus." *Psychonomic Science* 1 (1964): 47–8.

Whiten, Andrew, et al. "Culture in Chimpanzees." *Nature* 399 (1999): 682–85.

Whorf, Benjamin Lee. "Science and Linguistics." *Technology Review* 42 (1940): 227–31, 247–48. Reprinted in *Language, Thought, and Reality: Selected Writings of Benjamin Lee Whorf*, edited by John B. Carroll, 207–19. Cambridge, MA: Technology Press of MIT; New York: Wiley, 1956.

Wich, Serge A., et al. "A Case of Spontaneous Acquisition of a Human Sound by an Orang-utan." *Primates* 50 (2009): 56–64.

Wiessner, Polly W. "Embers of Society: Firelight Talk among the Ju/'huansi Bushmen." *Proceedings of the National Academy of Sciences* 111 (2014): 14,027–35.

< 249 >

Wilkins, Jane, Benjamin J. Schoville, Kyle S. Brown, and Michael Chazan. "Evidence for Early Hafted Hunting Technology." *Science* 338 (2012): 942–46.

Williams, George C. *Adaptation and Natural Selection: A Critique of Some Current Evolutionary Thought*. Princeton, NJ: Princeton University Press, 1966.

Williams, Justin H. G., Andrew Whiten, Thomas Suddendorf, and David I. Perrett. "Imitation, Mirror Neurons, and Autism." *Neuroscience and Biobehavioral Reviews* 25 (2001): 287–95.

Wilson, David S. *Darwin's Cathedral: Evolution, Religion, and the Nature of Society*. Chicago: University of Chicago Press, 2002.

Wray, Alison. "Dual Processing in Protolanguage: Performance without Competence." In *The Transition to Language*, edited by Alison Wray, 113–37. Oxford: Oxford University Press, 2002.

Wundt, Wilhelm. *Die Sprache*. 2 vols. Leipzig: Enghelman, 1900.

Wunn, Ina. "Beginning of Religion." *Numen* 47 (2000): 417–52.

Wynn, Thomas, Karenleigh A. Overmann, and Frederick L. Coolidge. "The False Dichotomy: A Refutation of the Neandertal Indistinguishability Claim." *Journal of Anthropological Sciences* 94 (2016): 1–22.

Xu, Jiang, et al. "Symbolic Gestures and Spoken Language Are Processed by a Common Neural System." *Proceedings of the National Academy of Sciences* 106 (2009): 20,664–69.

Zeshan, Ulrike. "Sign Language in Turkey: The Story of a Hidden Language." *Turkic Languages* 6 (2002): 229–74.

Zubrow, Ezra B. "The Demographic Modeling of Neanderthal Extinction." In *The Human Revolution: Behavioral and Biological Perspectives on the Origins of Modern Humans*, edited by Paul Mellars and Chris Stringer, 212–31. Princeton NJ: Princeton University Press, 1989.

< 250 >

Index

Acheulian, the, 185–86

Africa: and "big bang" theory, 35–36, 38, 51; environment of, 91; foraging in, 193; and language, 51–52, 180, 210n33; migration from, 29, 34–35, 37, 51, 167, 170, 201; predators in, 47, 91; and symbolic behavior, 38; and technology, 38, 201; tribes of, 106, 159

afterlife, belief in, 111–12

alarm calls, 175

Alexander, Richard D., 92

Andrews, Kristin, 70, 138

Animal Farm (Orwell), 7, 140

Animal Minds (Griffin), 70

Aquatic Ape, The (Morgan), 95

aquatic ape hypothesis, 48, 94–96, 162–63, 174

aquatic phase. *See* aquatic ape hypothesis; evolution

Arbib, Michael, 192, 218n16

arbitrariness, myth of, 152–54. *See also under* speech

Ardipithecus ramidus, 91, 141, 221n70

Ardipithecus kaddaba, 91

Armstrong, David F., 128

artificial intelligence, 58–59

Atkinson, Quentin, 51, 210n33

Auden, W. H., 67

Auel, Jean, 161, 162

babbling. *See under* speech

Babel, Tower of, 24–25, 30

bans on discussion of language origins, 40–41, 49. *See also* Linguistic Society of Paris; Philological Society of London

bear cults, 112–13

Bellugi, Ursula, 128

Bickerton, Derek, 30, 80

"big bang" theory, 27, 34, 36–38, 40–41, 47, 51, 53, 167, 169, 196, 199. *See also* Africa; Chomsky, Noam; evolution; genes; language evolution

Bindel, Julie, 118–19

bipedalism: and aquatic ape hypothesis, 95–96, 141; and early birth, 708; in early hominins, 91, 140; efficiency of, 141; facultative, 142; freeing the hands, 141–42; obligate, 142; and obstetrical dilemma, 97; origins of, 141, 221n70; as pain in the ass, 8

Bloom, Paul, 44–45, 47–48, 62

Boë, Louis-Jean, 165

Bogin, Barry, 98

bonobo: common ancestry with humans, 4, 6, 39, 47, 89, 91, 140–41; communication in, 136; empathy in, 86; extinction of, 137; future thinking in, 71; and gesture, 136–38; keyboard communication in, 136, 175, 192, 199; language in, 33; memory in, 71; and mind wandering, 63; pointing in, 91, 139, 175; vocalization in, 163

Borges, Jorge Luis, 11–12

Boyd, Brian, 94, 104, 110–11, 198

< 251 >

brain: and aquatic phase, 48, 97; asymmetry of, 8–9, 37; brain imaging, 50, 59, 63, 74; in Denisovans, 35, 37; and emotion, 87; expansion of, x, 1, 22, 30, 42–43, 48, 73, 92, 94, 142, 193; and experience, 32; fully modern, 36; and genes, 37, 46, 94, 151, 165–69, 215n31; and group size, 92; growth of, 7–8, 97; and language, 53; and memory, 59, 73–80; and mind wandering, 73–80; in Neandertals, 34, 37; and remapping, 77–78; and social intelligence, 94; wiring of, 27, 30. *See also* entorhinal cortex; FOXP2 gene; hippocampus

Brooks, Alison, 38
Brown-Schmidt, Sarah, 184
Browning, Robert, 67
Burling, Robbins, 147, 155
Byrne, Richard, 90, 137

caching: in birds, 66, 68–69; in chimpanzee, 71; in Clark's nutcracker, 66; in scrub jays, 69. *See also* memory
Call, Josep, 89
central pattern generator, 161
Chaminade, Thierry, 185
Chaser (border collie), 33, 136, 150–51, 175, 191, 227n7
Cheney, Dorothy L., 175
child development: adolescence, 98; childhood, 9, 98, 102; infancy, 6, 8, 32, 86, 97–98, 134–35, 139; juvenility, 98
chimpanzee: brain size of, 176; as "bush meat," 38; common ancestry with humans, 4, 6, 39, 91, 140–41; communication in, 89–90, 132, 136–37, 192; displacement in, 135; empathy in, 86; facial expression in, 155; and FOXP2 gene, 166; future thinking in, 71–72; genetic comparison with humans, 6, 166, 205n6; and gesture, 127, 136–38, 192, 199; internal thinking in, 102; knuckle walking in, 141; language in, 28, 47, 127, 138–40; laughter in, 157; memory in, 69–71; and mime, 137; myelination of brain,

99; pointing in, 9, 90, 135, 139, 220n43; spine of, 140; theory of mind in, 87–90; tool making by, 22, 138; vocalization in, 131–33

Chomsky, Noam: against natural selection, 32–33, 41; and animal communication, 131; archaeological support for, 34–36; "big bang" theory, 9, 48, 165, 169, 191; and discrete infinity, 9–10, 183; and evo-devo, 46; and externalization, 30; and I-language, 29–31, 33, 57, 61, 79–80, 189; and language as miracle, 29, 40; language as organ, 40, 53; language as program, 59; review of Skinner's *Verbal Behavior*, 11; and thought without language, 60; and unbounded Merge, 29–31, 185, 187–88, 199, 207n6; and universal grammar, 14–15, 30, 45–46, 48–49, 190

Christiansen, Morten, 1, 21, 46, 53
Clayton, Nicola, 69
cognitive linguistics, 21
cognitive niche, 48, 53, 92, 186, 193
Condillac, Abbé Étienne Bonnot de, 124
construction grammar. *See under* grammar
conventionalization, 5, 124, 144, 154, 170–71, 173–74, 176, 179–80, 182, 188, 194–96. *See also under* sign language; speech; words
cooperation, 86, 92–93, 193. *See also* theory of mind
Corballis, Michael C., 66–68, 128, 192, 218n16
Corkin, Suzanne, 74
Crawford, Michael A., 97
crime stories: dark side of, 116; role of detectives, 116–18; as morality tales, 116; and murder, 116–17, 120; origins of, 116; and science, 117. *See also* stories
Critchley, MacDonald, 126
Crow, Timothy J.: and brain asymmetry, 37; and handedness, 37; and language, 37; and speciation of *Homo sapiens*, 37, 47; and theory of mind, 37; and X and Y chromosomes, 37

< 252 >

Darwin, Charles: continuity between humans and animals, 79, 89, 176; "Darwin's Mistake," 28, 89, 207n2; on emotional expression, 85; on gestures, 130–31; and incremental change, 3, 23, 27–28, 40–42; and language evolution, 3, 40–41; and pre-adaptation, 42; on religion, 113–14; on sign language, 144, 158; theory of evolution, 1–3, 44

Dawkins, Richard, 28, 114

Deane, Paul, 183–84

death: acceptance of, 109; from choking, 164; and Maori legend, 108; and modern science, 109; and religion, 112

Dediu, Dan, 169, 210n33

default mode network, 63, 75. *See also* mind wandering

Denisovans: brain size of, 35, 37, 96, 170; common ancestry, 35, 169; distinct from humans, 37, 169, 200; extinction of, 36, 96; genetic make-up, 35, 37, 94, 169; mating with humans, 35, 37; the question of speech in, 37; sign language in, 146

Descartes, René, 26–27, 72

Descent of Man, The (Darwin), 41, 114

detective: as hero, 116–17; as problem-solver, 116–17; and science, 117; as voyeur, 118, 217n32

de Waal, Frans, 86, 136

directed scratch, 87, 90, 136–37. *See also* gesture

discrete infinity, 9, 11, 18–19, 29, 79, 183

displacement, 29–30, 61, 80, 189

docosahexaenoic acid (DHA), 97

Donald, Merlin, 57, 142

Dor, Daniel, 53–54, 61, 186, 189

"dual-stream" theory: dorsal stream, 151–52; of speech perception, 151; ventral stream, 151; of visual perception, 151; "what" vs. "where," 151. *See also under* speech perception

Duarte, Marcos, 97

Duff, Melissa, 184

Duffy, Carol Ann, 157

Dunbar, Robin, 92, 113, 156–57

Dunbar number, 92

Einstein, Albert, 64

Eldredge, Niles, 44

Elizabeth of Palatine, 26–27

empathy: in animals, 86; and fiction, 103–4; in infants, 86, 214n12; and mirror neurons, 108. *See also* theory of mind

entorhinal cortex: grid modules in, 78–79; hippocampal mapping, 77–78. *See also under* hippocampus

Evans, Nicholas, 16, 18, 50

Everett, Daniel, 53–54, 186, 202

evolution: of animals, 55, 80, 166; aquatic ape hypothesis, 94–96, 162–63; "big bang" theory, 27, 34, 36–38, 40–41, 47, 51, 53, 167, 169, 188, 196, 199; and bodily expression, 131, 163; of brain, 184; and creativity, 197; cultural, 53; of empathy, 86; and evo-devo, 46; and exaptation, 42; of the eye, 28; and fire, 107; and the future, 68; great leap forward, 34, 36–37, 170, 195; of the mind, 28, 32–33, 55, 79, 86, 93, 207n2; and natural selection, 1, 3, 22, 27–28, 31–33, 36, 40–53, 62, 114, 187–88, 200; of play, 104; of primates, 38–39, 48, 134, 148, 150; and punctuated equilibrium, 44; and religion, 113–14; and savanna, 95; and space, 191; and spandrels, 42; theory of, 1, 3, 27, 37, 39–40, 45, 113; as tinkerer, 43, 158; of tools, 53, 185, 200–201. *See also* language evolution

Evolution of Language, The (Fitch), 48

Expressions of the Emotions in Man and Animals (Darwin), 85, 144

facial movements: as bridge between gesture and speech, 155–57; in great apes, 155; in sign languages, 156; as social signals, 155–56; voluntary control of, 155. *See also under* sign language

Fano, Giorgio, 127, 128

Fauconnier, Gilles, 82

Faulkner, William, 18

< 253 >

fiction. *See* stories

fire, 106–7

Fitch, Tecumseh, 48–49, 165

Fodor, Jerry, 32, 57

FOXP2 gene: and Broca's area, 166; exon 7 region, 166–67; KE family, 165–66; and language, 167, 169; in mouse, 166, 226n75; mutation on, 166; 166–68, 170; and plasticity, 168; in songbirds, 167; and speech, 167–68, 170; as transcription factor, 168; vocalization and, 166–68. *See also* speech

Frame, Janet, 178

Fry, Stephen, 19

Fuller, John, 65

future thinking: in animals, 67–69, 71–75, 77, 79, 194; and brain asymmetry, 212n35; as extension of episodic memory, 67; imagining future episodes, 66, 77–78, 177, 184, 194; and language, 79, 171, 181–82; and mind wandering, 63; and planning, 55, 63, 102, 105, 142, 191, 194; in religion, 113. *See also* mental time travel; mind wandering

Gallaudet University, 126–28

Gell-Mann, Murray, 180

genes: and "big bang" theory, 37, 44, 91, 170; and brain size, 94, 215n31; Darwin's ignorance of, 3; shared with Denisovans, 37, 169; in great apes, 6; homeobox (Hox) genes, 45; in language evolution, 1, 37, 45, 151; and multicellularity, 198; number of, 45; shared with Neandertals, 37, 169; and psychosis, 37; and reading ability, 22; regulatory, 45; and the Rubicon, 203; and social evolution, 170; in songbirds, 151; and theory of mind, 37; and thought, 58. *See also* FOXP2 gene; Slit-Robo RhoGTPase-activating protein 2 (SRGAP2) gene

Genie, 6–7

Gentilucci, Maurizio, 157

gesture: in apes, 87, 90, 127; and cave drawings, 127; as communication, 84; facial, 8; as iconic, 142, 144, 152; in infants, 8; intentionality of, 90, 131, 135, 137–39; in Italians, 123–24; and mirror neurons, 128–30; as origin of language, 121, 123–46; as referential, 90; during speech, 8, 123–24; in stories, 106; voluntary control of, 131, 133

Gibbs, Wolcott, 180

Goodall, Jane, 131–32

Gould, Stephen Jay, 42–44

grammar: complexity of, 17, 52, 80, 202; construction grammar, 21; and episodes, 170; grammaticalization, 181–83; and inferior parietal lobule (IPL), 183–84; and mind wandering, 80; minimalism, 48, 188; rules of, 19; and space, 183–85, 190; theories of, 20–21, 29, 31; and tools, 184–86, 200; unbounded Merge. *See also* universal grammar

grasping, 129–30, 134, 141, 157–58, 184, 188, 191, 199

great leap forward. *See* evolution; evolution of language

Greene, Robert Lane, 5

Grice, Paul, 8–82, 90

Griffin, Donald R., 70

grooming, 87, 136, 155–56. *See also under* language evolution

Haldane, J. B. S., 160

hand-mouth connection, 147–48, 222n3

Harari, Yuvai Noah, 102

Hardy, Sir Alister, 95

Hebb, Donald, 104

Hewes, Gordon W., 127–28

Hickok, Gregory, 151–52

Higgs boson, 115, 117

hippocampus: asymmetry of, 184; in birds, 77, as cognitive map, 75–76, 184, 190, 199; and consolidation, 76; and discrete infinity, 79; and entorhinal cortex, 77–79; and fiction, 74, 79, 102; and language, 184; and memory, 73–76, 102, 184; and mental time travel, 74–76, 177; and place cells, 75–77; in the rat,

< 254 >

76–79; and recursion, 190, 199; scaling in, 77–79; in taxi drivers, 75. *See also* memory

hippocampus minor, 72–73

Hitchens, Christopher, 114

Hobaiter, Catherine, 90, 137

Hobbes, Thomas, 25–26, 31

Hockett, Charles, 161

Hoffecker, John, 36, 200

Holyoak, Keith, 89, 207n2

Homer, 109–10

Homo: the genus, 91, 94, 96, 98, 148, 157, 162–63, 176; *Homo erectus*, 38, 96, 102–3, 164; *Homo ergaster*, 142; *Homo habilis*, 96; *Homo narrans*, 102; *Homo sapiens*, 35, 37–38, 47, 91, 96, 98, 146, 160, 169, 185–86, 191–92, 200–201. *See also* Denisovans; Neandertals

Hopper, Paul, 183

Hubbard, Edward, 153

Humphrey, Nicholas, 93

Hurford, James, 20, 48, 173

Huxley, Thomas Henry: as Darwin's bull-dog, 71–73; and hippocampus minor, 72–73; vs. Owen, 73, 212n45

I-language, 29–33, 48, 57–58, 61, 79–80, 189. *See also* mind wandering; universal grammar

Iliad (Homer), 109

imageless thought, 60

incredulity, argument from, 28

instinct: and competition, 193; cooperation, 92; future survival, 68; language as,15, 41, 44, 207n18; and migration, 68; and monkey calls, 222n10; and sharing, 193

Instruction of Imagination, The (Dor), 53, 61

intentionality: in chimpanzees, 89; of gestures, 90, 131, 135, 137–39; and mirror system, 191; of pointing, 135; of tool use, 186; of vocalizations, 132–33, 154, 166, 175. *See also* gesture; mirror neurons

Jackendoff, Ray, 44–45, 59

James, Henry, 18, 198

James, William, 101, 112, 188, 196, 218n5

Jarvis, Erich, D., 133

Johansson, Sverker, 169

Kant, Immanuel, 57

Kanzi, 33, 136, 139–40, 151, 163, 175, 192, 199. *See also* bonobo

Karlsson, Fred, 14

Karvonen, Elin, 72

Kendon, Adam, 161, 224n53

Kingsley, Charles, 72

Kingsley, Mary, 159

Kinsler, Katherine, 32

Kirby, Simon, 1

Klein, Richard, 36

Klein, Stanley, 65, 67

Klima, Edward S., 128

Knott, Alistair, 188, 199

Kohler, Wolfgang, 153

Koko (gorilla), 136, 138–39

Kulpe, Oswald, 60

Kundera, Milan, 177

language: behaviorist approach to, 10–11, 21–22; as biological, 5, 9; connectionist theories of, 21; critical period in, 6–7; as cultural, 5; and discrete infinity, 9–10; and displacement, 55, 80, 130, 135, 172; and externalization, 30–31, 60, 80; faculty of language in the broad sense (FLB), 48–49; faculty of language in the narrow sense (FLN), 48–49; generativity, 4, 11, 14, 18, 20–21, 36, 79–80, 183, 190–91; hippocampus, role of, 184; left-brain specialization, 8; metaphors in, 176, 184, 202; as miracle, 24–39; and natural selection, 40–54; as organ, 40, 53; as ostensive-referential, 90, 192, 214n6; and recursion, 12–14, 20; as rule-governed, 9, 10; second-language learning, 7; as sharing, 45, 54–55, 61–62, 90, 105, 121, 138, 189, 193–94, 196, 198, 200; templates for, 18–20; as tool, 53, 186, 197–98, 202; underdeterminacy of, 84, 90, 192, 196; as uniquely human, 1, 3–6;

< 255 >

language (*continued*)
  universality of, 6, 14–18. *See also* Chomsky, Noam; language evolution; sentences; words
Language: The Cultural Tool (Everett), 53
language evolution: "big bang" theory, 27, 34, 36–38, 40–41, 47, 51, 53, 167, 169, 188, 191, 196, 199; and evo-devo, 46–47; and exaptation, 42–44; gestural origins, 9, 123–46; great leap forward, 34, 36–37, 170, 195; grooming as precursor, 136; as Just-So story, 202–3; and natural selection, 1, 3, 22, 27–28, 31–33, 36, 39–54, 62, 114, 187–88, 200–201; and punctuated equilibrium, 44; as spandrel, 42–44
Language Instinct, The (Pinker), 44
language-of-thought hypothesis (LOTH), 57
languages: !Xóõ, 51; Arabic, 25; Bininj Gun-Wok, 17; Cayuga, 17; Chinglish, 52; English, 4, 60; French, 52, 60; Greek, 52; Hebrew, 24; Iatmul, 14; Ilgar, 16; Italian, 16, 52; Latin, 16; loss of, 5; Mandarin Chinese, 4, 15; Nadëb, 180; Navajo, 197; Pirahä, 14, 202; Romance languages, 4; Russian, 15; Seneca, 16; Silbo Gomero (whistled language), 50; Spanish, 4; Thangmi, 17; Tiriyo, 16; Tobati, 180; Turkish, 16; Turkish Sign Language, 145; of Vanuatu, 5; Walpiri, 181; Warao, 180; Wik Ngathana, 180. *See also* sign languages
larynx: descent of, 164; in speech, 148, 156, 164, 168; as a valve, 43
laughter: as emotional expression, 85; epidemics of, 157; faking of, 157; as preliminary to speech, 156–57; suppression of, 132
Leavens, David, 138
Levinson, Stephen, 17–18, 169
Lewontin, Richard C., 42
Liberman, Alvin, 148–49
"Library of Babel, The" (Borges), 11–12, 14
Lieberman, Daniel, 165
Lieberman, Philip, 164, 165

Linguistic Society of Paris, 40, 49
lipreading, 156. *See also* speech; words
Locke, John, 25, 57
Locke, John L., 98
Loftus, Elizabeth, 66
logic: and religion, 26; and space, 184; vs. stories, 188
Lyn, Heidi, 135

Macaulay, Thomas Babbington, 32, 207n13
MacDonald, John, 156
Machiavelli, Niccolo, 92
Machiavellian loop, 92
MacNeilage, Peter, 160
magic, belief in, 107, 111–13, 115. *See also* religion
Maxim of Quantity, 82. *See also* Grice, Paul
McBrearty, Sally, 38
McBride, Glen, 142–43
McCarthy, Robert, 164
McGurk, Harry, 156
McNeill, David, 148, 161
Mellars, Sir Paul, 36
memory: auditory memory, 150, 222n11; autobiographical memory, 63, 102; capacity of, 176–78; and concept of self, 65, 75; for concepts, 176; consolidation of, 76; episodic memory, 65–66, 73–74, 102, 178; false memories, 66; of Google, 59; and hippocampus, 73–76, 102, 177, 184; and imagining the future, 66–68, 71, 74, 79; and language, 150, 188; limitations of, 19–20; and mental time travel, 66–67; and mind wandering, 65; semantic memory, 65–66, 74; for sounds, 150; for stock phrases, 19–20; whether uniquely human, 66–72; working memory, 53. *See also* mental time travel
mental time travel: in animals, 66–75; extension of episodic memory, 66–67; and hippocampus, 73–78, 177; and recursion, 43, 55; and stories, 56; whether uniquely human, 33, 67–72, 76–78. *See also* future thinking; memory
Meunier, Helène, 134

< 256 >

Milne, A. A., 97
mime: in apes, 137–39; and brain, 143;
    conventionalization of, 194; in cross-
    language communication, 142; in early
    Pleistocene, 194; holistic nature of, 179;
    and language, 154, 173; mime artists, 106,
    127, 142; mimetic culture, 142, 176; and
    mirror system, 130; as origin of sign lan-
    guage, 5, 18, 127–28, 143–46; and speech,
    158, 161, 171; and stories, 106, 142–43, 194;
    and tools, 185; and words, 174, 179. *See
    also* language evolution; mirror neurons
mindfulness, 65
mind reading. *See* theory of mind
mind wandering: in animals, 69–72, 75–77,
    82; and creativity, 64; and default-mode
    network, 63, 75; and dreaming, 64, and
    fantasy, 63; and fiction, 63; and hippo-
    campus, 76–77; and I-language, 80;
    memory, 65; negative aspects, 64–65.
    See also default mode network; mental
    time travel
miniaturization, 158, 195. *See also* speech
minimalism. *See under* grammar
mirror neurons: and area F5, 129–30; aut-
    ism, 128; critique of, 152; and displace-
    ment, 130; and dorsal system, 152; and
    empathy, 128; and FOXP2 gene, 166;
    gestural theory, 128; and grasping, 129;
    in human brain, 130; and imitation, 128;
    and intentionality, 191; and language,
    129–30; and language areas (Broca's and
    Wernicke's areas), 129–30; as learned,
    22n9; and mime, 130; and motor theory
    of speech perception, 149–50; and
    schizophrenia, 128; and sound, 150; and
    symbols, 143; and vocalization, 150
mirror system, 129–30, 143, 150, 152, 166, 191.
    See also mirror neurons
modernity, 34, 38, 51, 201
Molaison, Henry (H.M.), 74
Morgan, Elaine, 95
Moser, Edvard I., 76–77, 79, 213n54
Moser, May-Britt, 76–77, 79, 213n54
Müller, Friedrich Max, 3–4, 40

natural selection: *See under* evolution
Neandertals: brain size of, 34, 96; common
    ancestry with, 34, 169; as distinct from
    humans, 36, 169–71, 226n82; extinction
    of, 35–36, 96, 160, 200; and FOXP2 gene,
    166–67; genetic make-up, 35, 37, 94;
    and hunting technology, 38, 185; mating
    with humans, 35, 37; the question of sign
    language in, 146, 163–64; the question of
    speech in, 164–71
Nelson, Katherine, 62, 102
Newton, Isaac, 22, 207n40
Nietzsche, Frederich, 125
Nikolaïdes, Kimon, 127
Niles, John, 102

Oatley, Keith, 103–4
Odyssey (Homer), 110
O'Keefe, John, 184, 213n51
Oldowan, The, 185–86
On the Origin of Species (Darwin), 1
Origins of the Human Mind (Donald), 142
Orwell, George, 7, 102, 14
ostensive-referential. *See under* language
Osvath, Mathias, 72
Owen, Richard, 72–73

Paley, William, 28
Pandya, Deepak, N., 151
pantomime. *See* mime
Penn, Derek, 89, 207n2
Petkov, Christopher, 133
Petrides, Michael, 151
Pfeiffer, John E., 169
Philological Society of London, 40
pidgins, 182–83
Pinker, Steven, 31–32, 44–45, 47–48, 57,
    61–62, 131, 153, 173, 178, 206n16, 226n5
Plato, 145
play: in adults, 105; in animals, 104; in birds,
    104; in children, 104–5; chimpanzees,
    90; play acting, 101; play-fighting, 104;
    as stories, 103–4
Pleistocene: and aquatic features, 96; and
    brain size, 48, 94, 99, 176, 193; and car-

< 257 >

Pleistocene (*continued*)
nivorous animals, 91; and coastal habitat, 96; dangerous predators, of, 193; dating of, 48, 91; and genus *Homo*, 94, 98; and language, 186, 193; and mime, 194; and social intelligence, 99; and tools, 184–86
Plutarch, 189
Poeppel, David, 151
pointing: as communication, 85; in infants, 8–9, 90, 134; as precursor to language, 134–35; in primates, 9, 80, 134–36, 140, 163; and requesting, 140; right-hand preference in, 134; and sharing, 139; universality of, 124, 217n1. *See also* gesture
Pollick, Amy, 136
polysemy, 61. *See also* words
Povinelli, Daniel, 88–89, 207n2
Premack, David, 28, 87, 89, 93–94, 132
Prometheus, 30, 44, 49, 53, 166–67
Provine, Robert, 157
punctuated evolution. *See under* evolution

Question of Animal Awareness, The (Griffin), 70
Quilliam, Susan, 119
Quintilian, 121

Racine, Timothy, 138
Ramachandran, Vilayanur, 128, 153–54
Rand, David, G., 93
Rankin, Ian, 117–18
Reagan, Ronald, 65
religion: adaptiveness of, 114; dark side of, 114; and Neolithic revolution, 114; and recursion, 114; vs. science, 113–15. *See also* stories
ribena, 10
Rico (border collie), 33, 136, 151, 175, 191, 227n7
Roebroeks, Wil, 169
romantic fiction: and feminism, 119; formula of, 118; popularity of, 118–19, 217n36; as rape fantasy, 118; varieties of, 119. *See also* stories

roundworms, 46, 209n16
Rousseau, Jean-Jacques, 124–25, 127–28
Rovelli, Carlo, 64
Rowland, David C., 76–77, 79
Rubicon, the, 3, 23, 28, 30, 80, 112, 155, 189, 200, 203
Ruhlen, Merritt, 180
Russon, Anne, 70, 138

Sassoon, Siegfried, 61
Saussure, Ferdinand de, 152
Scott-Phillips, Thom, 84, 90, 192, 214n6
scrambling languages, 181. *See also* languages
second-language learning, 4, 7. *See also* languages
"Secret Life of Walter Mitty" (Thurber), 63–64
sentences: ambiguity of, 84; connectionist modelling of, 21; and context, 84; data oriented parsing model (DOP), 20; discrete infinity and, 11, 19; embedding of phrases in, 12–14; and episodes, 177–78, 187; function words in, 174; generation of, 4, 10, 36; grammar of, 21–22, 59; grammaticalization in, 181–83; length of, 18–19; psychological limits on, 13, 18, 185; recursion in, 19, 21; rules for, 11, 20; single-word sentences, 16; storage of, 20; and stories, 198–99; translation of, 197; unbounded Merge, 29; word order in, 180–81. *See also* grammar; languages
serial founder effect, 201
Seyfarth, Robert M., 175
Shaw, George Bernard, 5
sign language: in African tribes, 159; as agglutinative, 18; antiquity of, 145; autonomy of, 197; brain activity in, 143; conventionalization in, 144–45, 154; 171, 179; in the deaf, 106; and Deaf culture, 159; diversity of, 5; ease of learning, 163; facial movements in, 155; as Gallaudet University, 126; and gesture, 8, 163; in great apes, 6, 48, 133, 136; in hunter-gatherers,159; iconic elements in, 144,

< 258 >

154; in Neandertals, 146, 163–64; as origin of language, 126–28, 145–46; physiological costs of, 158; rapid emergence of, 145–46; signed stories, 106; as symbolic, 172; as true language, 125–26, 128; word order in, 180; 179–80, 197, 199, 229n. *See also* mime

sign languages: Al-Sayyid Bedouin Sign Language, 146; American Sign language, 126, 144; Nicaraguan Sign Language, 145–46, 179. *See also* languages

Simon, Herbert, 58

Skinner, Burrhus Frederick, x, 10–11, 21–22, 68

Skyrms, Brian, 92

Slit-Robo RhoGTPase-activating protein 2 (SRGAP2) gene, 94, 215n31. *See also* genes

Slobin, Dan, 62

Slocombe, Katie, 89, 132

social intelligence, 91–94, 97, 99

Socrates, 145

Speaking Our Minds (Scott-Phillips), 84, 192

spears, 34, 38, 138, 208n33

speech: and animal vocalizations, 130; and aquatic ape hypothesis, 162; arbitrariness, myth of, 152, 162, 171; babbling, 8, 41; brain areas in, 143; and breathing, 43; Broca's aphasia, 151; and choking, 164; and conventionalization, 154, 170, 195; and descent of larynx, 164; dominance of, 125, 16; emergence of, 121, 124; as energy efficient, 158, 171, 195, 229n6; evolution of, 41; and FOXP2 mutation, 165–69; and freeing of the eyes, 158–59; and freeing of the hands, 158, 171; gestural accompaniment, 124–26, 148, 157, 161, 195; as gesture, 148, 152, 156; in great apes, 127; iconic elements of, 152–54; as invention, 195; and laughter, 156–57; vs. mime, 127; as miniaturization, 158, 195; mutation of, 171; and Neandertals, 37, 163–64, 167–70; origins of, 144; in parrots, 41; parts of, 31, 179, 182, 187; and

pedagogy, 170; phonemes, 50–51; rhythmicity of, 161; vs. sign language, 124–25; and social status, 51; structure of, 18, 29; subvocal, 57; and vocal tract, 165; and voluntary control, 133, 157, 162. *See also* language; speech perception

speech perception: by apes, 33, 150; by babies, 151; by computers, 149; by dogs, 33, 150, 175; dual-stream theory, 151; and fast mapping, 175; lipreading, 156; and mirror neurons, 149–50, 152; motor theory of, 148–49, 223n15. *See also* speech

Spelke, Elizabeth, 32

Spenser, Edmund, 18

Sperber, Dan, 82

Sterelny, Kim, 93

Stokoe, William C., 128

storied thoughts, 62, 187

stories: and animals, 112–13; children's stories, 104–5, 112–13, 120; crime stories, 104, 115–18; as distinctively human, 29, 62, 100, 102, 131, 198; epic tales, 108–10; and episodes, 171, 187; and fantasy, 103–5, 117–19; and fire, 106; and folklore, 107–8; and future adaptation, 104–5; and gesture, 106; Gilgamesh, Epic of, 108–9; and hunter-gatherers, 106, 109, 142–43, 194; and imagination, 103, 196; and indigenous Australians, 107; and language, 56, 172, 196, 198, 200; and Maori, 107–8; media of, 56, 62, 107; and mental time travel, 101–2; mimed, 142, 194; as mode of thinking, 101, 187–88; during night, 106–7; as play, 104–5, 196; and religion, 110–15, 199; romantic fiction, 118–19; and the self, 102; and social reform, 119–20; and social world, 103; structure of, 199; superhuman elements, 109–12, 120; and theory of mind, 101, 103–4; time sequence in, 102

Stout, Dietrich, 185

Suddendorf, Thomas, 66–68, 207n2

Sutherland, John, 108, 110

Syntactic Structures (Chomsky), x, 11, 21

Szathmàry, Eörs, 198

Tattersall, Ian, 36, 185, 203
Tennyson, Alfred, Lord, 193
theory of mind: in animals, 85–91; in children, 98; and cognitive niche, 9, 193; and cooperation, 92–93, 193; and fiction, 103–4; and genetic change, 37; and hunting, 92; and language, 56, 81–85, 192; and magic, 112–13; and recursion, 43, 93, 113; and religion, 113; and social intelligence, 91–94
Thurber, James, 63
Time magazine, 80, 214n4
Tobias, Philip, 95, 162
Tomasello, Michael, 18, 89, 135, 138–39, 145–46, 192, 206n27
Traugott, Elizabeth, 183
Tulving, Endel, 65–67
Turin, Mark, 5, 17
Turing, Alan, 58–59, 211n7
Turing test, the, 58–59

unbounded Merge, 29–31, 43, 49, 139, 185, 187–88, 190, 199, 207n6, 228n33. *See also* Chomsky, Noam; grammar; sentences
underdeterminacy. *See under* languages
universal grammar, 14–18, 20–21, 29–31, 45–46, 48–49, 166, 190, 202, 206n27. *See also* Chomsky, Noam; grammar; I-language

Verbal Behavior (Skinner), x, 10–11, 21
Verhaegen, Mark, 95–96, 162, 164
Villa, Paola, 169
vocal learning, 131
Volterra, Virginia, 134
Vrba, Elizabeth S., 42
Vygotsky, Lev, 62, 105

Wallace, Alfred Russel, 27–28
Water Babies, The (Charles Kingsley), 72
Watson, John B., 57
Wearing, Clive, 73–75

Wearing, Deborah, 74
West, Mae, 87
Westenhöfer, Max, 94
Whitehead, Alfred North, 10
Whorf, Benjamin Lee, 173
Whorfian hypothesis, 173
Wilcox, Sherman, 128
Wild children: in fiction, 6; Genie, 6–7
Williams, George C., 45
Wilson, David Sloan, 114
Wilson, Deirdre, 82
Woodruff, Guy, 87, 89
words: abstract vs. concrete, 176, 187; in animals, 33, 150, 175, 227n8; as arbitrary (or not), 33, 152–56, 171, 195; articulation of, 25, 50, 148–49; and baby talk, 8; categories of, 16, 18, 45, 174, 176–77, 179; computer recognition of, 149; and conventionalization, 195; and displacement, 33; diversity of, 196, 201; emergence of, 134; fast mapping, 175; frequency of, 195; function words, 181–82; and gestures, 160; iconic component, 142, 144–45, 152, 154, 161, 173, 187, 194–96; inflections of, 52, 182, 197; lipreading of, 156; mapping to concepts, 15–18, 26, 45, 52, 58–59, 143, 172–74; memory for, 150; merging of, 29; multiple meanings of, 61; in parrots, 6; signs as, 8, 123; store of, 21, 30, 32, 150, 175–76, 201; stringing of, 9–11, 13, 30, 45, 178, 180–81, 190, 197; structure of, 226n1; as symbols, 143, 227n8; and thought, 57–59, 62. *See also* language
Wray, Alison, 19
Wundt, Wilhelm, 126, 218n5
Wurzburg School, 60

X and Y chromosomes. *See* Crow, Timothy J.

Zuberbühler, Klaus, 89, 132
Zubrow, Ezra, 160

< 260 >